PLANTS OF THE BIBLE

PLANTS OF THE BIBLE

AND HOW TO GROW THEM

ALLAN A. SWENSON

DRAWINGS BY MARTHA LIMBO HEATH

A CITADEL PRESS BOOK
PUBLISHED BY CAROL PUBLISHING GROUP

All Biblical quotations, unless otherwise noted, are taken from the King James Version of the Bible.

Designs for creating gardens of Biblical plants—herbs, fruits, and trees—as part of your home landscape were redrawn by Martha Limbo Heath from sketches prepared by landscape horticulturalist Peter Jon Swenson for this book.

Photos have been provided courtesy of W. Atlee Burpee Company; White Flower Farm, Litchfield, Connecticut; Neot Kedumim, Ltd., Kiryat Ono, Israel; and others as indicated, and are used by permission.

This book was originally published under the title *Your Biblical Garden*.

Copyright © 1995 by Allan A. Swenson

A Citadel Press Book
Published by Carol Publishing Group
Citadel Press is a registered trademark of Carol Communications, Inc.
Editorial Offices: 600 Madison Avenue, New York, N.Y. 10022
Sales and Distribution Offices: 120 Enterprise Avenue, Secaucus, N.J. 07094
In Canada: Canadian Manda Group, P.O. Box 920, Station U, Toronto, Ontario M8Z 5P9
Queries regarding rights and permissions should be addressed to Carol Publishing Group, 600 Madison Avenue, New York, N.Y. 10022

Carol Publishing Group books are available at special discounts for bulk purchases, for sales promotions, fund-raising, or educational purposes. Special editions can be created to specifications. For details, contact Special Sales Department, Carol Publishing Group, 120 Enterprise Avenue, Secaucus, N.J. 07094

Manufactured in the United States of America
10 9 8 7 6 5 4 3 2 1

Library of Congress Cataloging-in-Publication Data

Swenson, Allan A.
    Plants of the Bible : and how to grow them / by Allan A. Swenson.
        p. cm.
    "A Citadel Press book."
    Includes bibliographical references (p.   ) and index.
    ISBN 0-8065-1615-1 (pbk.)
    1. Plants in the Bible.   2. Gardening.   I. Title.
[SB454.3.B52S88   1995]
635—dc20
                                                        94–44950
                                                        CIP

To Lamar Robinson, a man of quiet courage, sound advice, common sense, loyal friendship and abiding faith.

Our world needs many more like him.

And, to my late Aunt Martha Dugdale, who took me to Sunday School regularly and focused me on the goodness there is in people, in life and in oneself if one perseveres.

ACKNOWLEDGMENTS

Thanks are due hundreds of people who have devoted tireless efforts in studying the scriptures and in providing helpful guidance in the writing of this book. Special thanks to the Reverend Lamar Robinson for his quiet counsel, Sarah Larkin Loening for her dedicated effort in founding the Biblical Gardens in New York; to friends—Protestant, Catholic, Jewish—who offered ideas and suggestions. May your gardens grow abundantly fruitful and reward you with their beauty for years to come.

PHOTO AND ART ACKNOWLEDGMENTS
With thanks for their help during research and in providing pictures, color photos, drawings and art.

Peter J. Swenson

D. R. Sabaka

W. Atlee Burpee & Co.

Park Seeds

J. Drayton Hastie

Magnolia Gardens

Neot Kedumim

Nogah Hareuveni

Paul Steinfeld

Netherlands Bulb Growers

CONTENTS

PROLOGUE

Whether you are a veteran gardener, looking for new growing horizons, or a beginner, itching to exercise and cultivate the growing talents you feel within yourself, a garden of Biblical plants can be a most rewarding and fulfilling experience.

In the beginning, God created the earth and the waters; and then He created the plants on the third day.

And God said, as we read in Genesis 1:11–12:

"Let the earth bring forth grass, the herb yielding seed, and the fruit tree yielding fruit after his kind, whose seed is in itself, upon the earth: and it was so. And the earth brought forth grass, and herb yielding seed after his kind, and the tree yielding fruit, whose seed was in itself, after his kind: and God saw that it was good."

When the work of creation was nearly done, God created man in his own image, male and female, and blessed them. On the sixth day, according to the scriptures in Genesis 1:29, 31:

"And God said, Behold, I have given you every herb bearing seed, which is upon the face of all the earth, and every tree, in the which is the fruit of a tree yielding seed; to you it shall be for meat . . .

"And God saw every thing that He had made, and, behold, it was very good. And the evening and the morning were the sixth day."

As you read the marvelous words in Genesis, describing the creation, you find the earliest references in the scriptures to plants of the Bible.

On the seventh day, God ended his work and rested.

"And every plant of the field before it was in the earth, and every herb of the field before it grew: for the Lord God had not caused it to rain upon the earth, and there was not a man to till the ground. But there went up a mist from the earth, and watered the whole face of the ground. And the Lord God formed man of the dust of the ground, and breathed into his nostrils the breath of lite; and man became a living soul." (Genesis 2:5–7)

"And the Lord God planted a garden eastward in Eden; and there he put the man whom he had formed." (Genesis 2:8)

"And out of the ground made the Lord God to grow every tree that is pleasant to the sight, and good for food; the tree of life also in the midst of the garden, and the tree of knowledge of good and evil." (Genesis 2:9)

"And the Lord God took the man, and put him into the garden of Eden to dress it and to keep it." (Genesis 2:15)

"And the Lord God caused a deep sleep to fall upon Adam, and he slept: and he took one of his ribs, and closed up the flesh instead thereof; And the rib, which the Lord God had taken from man, made he a woman, and brought her unto the man." (Genesis 2:21–22)

Further, in the story of the beginning, in Genesis, we read of the banishment of Adam and Eve from the Garden of Eden. For having eaten the forbidden fruit, God chastised them, sending them forth from the garden.

"Therefore the Lord God sent him forth from the garden of Eden, to till the ground from whence he was taken." (Genesis 3:23)

From the beginning, men and women have tilled the land, cultivating the earth to bring forth the flowers, the food and fruit, the herbs and trees that would sustain the generations to come.

Through the ages, men and women have found both food and beauty among the plants of our earth. This book is respectfully dedicated with grateful thanks to all who have helped cultivate the good earth to make it bloom and bear. Today, we too enjoy the glories of the flowers and the fruitful abundance from trees, vines, and other plants.

As you embark on new growing experiences with plants of the Bible, planting seeds, tending your plants, savoring their goodness and beauty, it is well to remember that what we have and are is a gift of God.

As you sow your seeds and plant your trees, you too are a part of the continuing creation of life and beauty. We are all, in fact, gardening with God.

ALLAN A. SWENSON
Kennebunk, Maine

INTRODUCTION

Plants of the Bible are alive and thriving today, perhaps much closer to home than you may realize. You may pass some of them on your way to work, or even grow some in your own home gardens without relating to their deeply rooted heritage in the pages of the scriptures. Over the centuries, in the gardens of religious orders as well as local churchyards, plants of the Bible have been lovingly tended. Today, you too can cultivate the blooming beauty and enjoy the fragrance of many of these plants, wherever you live. It is easier than you imagine, and the pleasures you reap can be as abundantly rewarding as the reading of the scriptures.

The Bible abounds in references to plants, as all who have read it have realized. From the earliest descriptions of the Garden of Eden, through the books of the Old and New Testaments, we can read many passages depicting the flowers and trees, the foods and herbs as they were in the Holy Land before and during the life of Christ. Since then, many plants of the Bible have been transplanted by pilgrims and travelers to the far reaches of our planet. As you read the Bible and the many references to the plants that grew there, you may wonder, as millions have, whether you could cultivate some of these plants yourself. The answer is yes. Not only can you grow some of them, but you can create your own Biblical gardens containing many more types of plants than you might expect. You can grow fruits and vegetables, herbs and spices, shrubs and trees, and even some plants indoors for their beauty, fragrance, and the pleasure that they give you from their special significance as plants of the Bible.

During the past decade, millions of people like you have discovered the pleasures and rewards of home gardening. From backyard plots to porch, patio, and even container gardens on rooftops in the cities, a growing phenomenon has continued to sprout and take root. Today, gardening is acknowledged as America's most popular family hobby.

With this popularity has come increased interest in plants of special significance and appeal. There are legions of avid orchid growers, throngs of cacti enthusiasts, countless numbers of carnivorous plant cultivators. Old-fashioned roses are regaining admirers, and a fond feeling of nostalgia has focused greater and well-deserved attention on the merits of many old-time garden favorites. Houseplants of graceful shape and form adorn homes, offices, restaurants, and shopping malls.

It comes as no surprise that Biblical plants also are sharing in this rebirth of interest in plants of special significance. In individual home gardens and among church members, renewed attention is being paid to Biblical plants. That is a welcome sign. Fortunately, plants that had their ancient roots thousands of years ago in the Holy Land do survive and thrive today all across America and around the world.

You may wonder how and why plants of the Bible have spread so far beyond the land of the Bible. For centuries before the birth of Christ and for centuries afterward, the Holy Land was a crossroads for many people. Some came for reasons of trade, others for reasons of war. All left their mark of change to greater or lesser degree.

From careful research through the works of Biblical and botanical scholars while developing this book, tracing the roots of the plants of the Bible, it was readily apparent that plants migrate with people. As people, for trade or other purposes, visited the land of the Bible, seeds and plants arrived with them. It is important to understand that caravans from many different lands traveled extensively through the regions of the Holy Land long before the time of Christ and even before much of the Bible was committed to writing.

Plants native to India, Africa, parts of Asia, and much of Europe were carried on caravans across hundreds upon hundreds of miles. Orchards and vineyards, grain fields and vegetable gardens were planted with these seeds, saplings, and rootstocks that were native to areas other than the Holy Land itself. As peoples moved and warred and traded, it was natural that this interchange of plant varieties would take place. Some of the plants mentioned in the Bible, therefore, may not necessarily be native to that area. For the purposes of this book, however, and based on the extensive work of both Biblical and botanical scholars, plants that existed in the Holy Land, whatever their origins, whether native or transplanted long ago, are considered if they are mentioned in the scriptures.

Considering that many plants from other parts of the world found their way to the Holy Land, it follows that plants of the Bible also found their way to other parts of the world. Some were moved in the reverse direction along the same early caravan routes of trade. Pilgrims also accounted for substantial movement, carefully digging up bulbs, saplings, and roots as well as gathering seeds in season that they could plant upon return to their native lands. The Romans conducted trade throughout their Empire, and many plants found in the Holy Land were favored for their beauty and fragrance as well as commercial values. These, too, were transported back to Rome and cultivated in other parts of the Roman Empire.

During the centuries that followed, especially during the time of the Crusades, plants that captured the attention and captivated the eye of all who traveled in the lands of the Bible were picked for return to distant countries from which the Crusaders, pilgrims, and merchants had come.

Today, the land of the Bible blooms and bears abundantly again. Many of the wild and domesticated plants that trace their true heritage through the pages of the Bible have many relatives around the world. And, today, more enthusiastic gardeners are turning their attention to the more exotic plants with historic importance as they expand their growing activities.

The first settlers to the United States also brought with them seeds and plants with which they could begin farming the land. Through the years, seeds and rootstocks from many countries arrived and were transplanted along with the migrations of diverse nationalities that became America. Among these plants were many that traced their origins back to the land of the Bible.

Some of the plants of the Bible that are available today are direct descendants of the fruits, flowers, vegetables, herbs, trees, and other plants mentioned in the Bible and resemble their forebears quite closely. Naturally, over the centuries, some have changed slightly as they have become acclimated to climates and growing conditions different from their native habitats.

Others sprouting from the same basic families are strikingly similar in appearance to plants that grew in Biblical days. They too trace their roots to the scriptures. Over the intervening centuries, however, plant breeders in many countries have utilized the rich sources of botanical specimens, some wild, many domesticated, from among the flower and food-crop plants that grew in the land of the Bible. From these sources and others, through careful crossbreeding, they have developed improved new varieties which have more suitable cultural characteristics and other desired traits than the original parent stock which they resemble. This selective breeding has contributed to the wealth of improved vegetables, hardy herbs, and high-yielding fruit trees which are available to us today.

As you read the Bible, you'll find that the scriptures from both Old and New Testaments abound in references to plants of many types. There are confusing variations between different versions of the Bibles as well as confusion created during translations of these different Bibles into different languages. That's understandable, for several reasons. Those who wrote the Bible originally, committing historical, religious, moral, and ethical lessons to print, were certainly not trained botanists. There was, in those days, no botanical and horticultural science. Botanical classification of plants is, in fact, a relatively modern science.

As you read further in the Bible, you may be puzzled by the discrepancies, seeming contradictions, misidentifications, and other inaccuracies as the differing versions refer to plants. Translation, spelling, and printing errors aside, keep in mind the original intent of the scriptures. The Old Testament originated in the form of songs, ballads, and poetry, as well as folk songs, handed down from generation to generation by word-of-mouth tradition. The moral and religious messages were of far greater importance than accurate plant identifications. The Bible was never intended by its writers, translators, or printers to serve as a botanical or even natural science text or reference.

Among Biblical scholars and botanists alike, today as in years ago, debate continues concerning the true identity of certain specific flowers, fruits, trees, and other plants mentioned in the scriptures. Consider another fact, that there have been through the ages dozens of different Bibles. It is understandable that there remains some confusion as to which plants really are which. Even the widely accepted Authorized Version of the King James Bible contains some misidentification of plants between the Biblical ones and common English ones. For example, aspens have been called mulberries and mulberries called sycamores. In other cases the plane tree became a chestnut and the apricot an apple.

No doubt these errors, whatever their cause, have created confusion. No doubt debates will continue as other scholars add their interpretations to the meanings of the scriptural passages as they pertain to certain plants of the Bible.

Precisely identifying every plant remains an illusive goal. Debate still rages most strongly concerning two of the best known plants of the Bible, the lily and the rose. Scholars have variously identified the lily of the valley in The Song of Solomon as a hyacinth, narcissus, or lily. Artists during the Middle Ages and the Renaissance often depicted the Madonna Lily with the Annunciation and Resurrection. Other authorities vote for the sternbergia, which most likely did not exist in Palestine. Most authorities, however, concur that, based on botanical factors, the lily of the valley and lilies of the field are more likely references to the Palestine anemone.

The rose perhaps has created more argument. Various authorities have offered the rose as a narcissus, anemone, tulip, or crocus. Digging into antiquity, through Biblical translations from the ancient Hebrew, authorities may not agree on what the Rose of Sharon actually was, except to say that it was not a rose as we know a rose today.

Despite these debates, and the lack of exact definition of some plants, most have actually been rather clearly and accurately identified. This has been made possible by careful study of the old scriptures and translations, and comparisons of the plants that are native today in the particular climates and soils of the Holy Land. By studying botanical nomenclature, tracing plant families and plant structure, and comparing cultural needs, today we can be reasonably specific in identifying the majority of the plants of the Bible.

Where there remains doubt or serious debate, you'll find in this book several logical determinations and the reasons for the identifications. In any event, you'll also find the cultural methods for the various plants, whether you choose to believe the lily is a hyacinth, lily, or anemone. In the various chapters, you'll also find scriptural references to which you can refer, as well as thoughts and observations of leading scholars.

The purpose of this book is to enable you to learn more about your favorite plants of the Bible and how to grow them successfully. It is not intended to be a definitive botanical text on which plants are which; rather, it is intended as a perspective on the plants of the Bible in a broader sense.

You'll find tips and advice both for growing crocuses outdoors and for forcing them into colorful bloom as indoor displays as well. You'll discover secrets for producing abundant fruit from apricot and apple trees, as well as learn new lessons in growing grapes, which are so frequently mentioned in the scriptures.

Throughout the pages of the Bible, you'll find references to herbs and vegetables. You'll be able to apply the tips and advice in this book to grow many of these plants more productively in your own gardens, and in containers if you desire.

In addition to the foods of the Bible, the food for the body, you'll find references to the beauty of the world of flowers, then and now. Those who wrote the Bible saw about them the beauty of the earth and the fulfillment obtained from the world in which they lived. You too can find rewarding and satisfying experiences growing the flowers of the Bible and enjoy the blooming beauty that these flowers can provide today.

For those who wish to explore the fascinating, growing world of plants of the Bible, there are sources for seeds and plants included in this book. From these sources, you can obtain the seed packets, roots, and plants as well as trees.

If you are a parent or a teacher, or participate in Bible study or teach Sunday School classes, you may wish to grow some of the easiest plants. Since some of the actual species and varieties do not respond to cultivation here, or are not generally available, you'll find plants listed which belong to the same family, are available, do resemble the Biblical plants, and can be grown successfully.

In addition, for your further study, you'll find details about some of the best Biblical gardens in America that you can visit, perhaps even participating in the activities of the organizations that have planted and cultivated these gardens.

In the past decade, millions more gardens have sprouted across the country as individuals and families have begun discovering the joy and satisfaction that comes with gardening. Around homes, in backyards, in apartments, offices, schools, more people like yourself are turning the good earth and tending their houseplants too, and enjoying God's beauty and abundance in their lives.

Whatever your own particular faith, you too can join in this growing adventure, planting and cultivating plants of the Bible. As you do, you'll find new meaning in the scriptures and the growing world around you.

As you grow your own Biblical plants, you also can share your experiences with friends, relatives, neighbors, and perhaps also help youngsters to gain an appreciation of God's living world which we all share.

In this book you will perhaps find food for thought as you grow vegetables and fruit for tastier eating. As you cultivate flowers and trees, you will also find new beauty and a growing realization that the miracles of plants can be an uplifting force for beauty and fulfillment in your life and the lives of those around you.

PART ONE

CHAPTER I

PLANTS YOU CAN GROW

Throughout the Bible, the scriptures are alive with colorful descriptions of plants and many other passing references to them. Unless you have acres of land it is almost impossible to grow them all. Some require rather specialized cultural conditions. Others are represented today by somewhat different versions of the original parent plants, judging from the studies of botanists who have traced the roots of such plants back to their original sources.

However, even if you have limited space, there are many that can become part of your home plantscapes, or be added to your indoor plant collections. The list of plants in this chapter provides a good starting point for selecting those that you may wish to grow. Since most gardeners at times become carried away with enthusiasm, a word of caution is in order. It is best to try a few, and expand your gardens of Biblical plants gradually. In that way, you can get to know each plant better and learn its cultural requirements to grow it satisfactorily. As your efforts reward you with success in abundance, you can increase your collection month by month and year by year.

Current botanical nomenclature is, of course, a comparatively recent scientific specialty. Even today, authorities occasionally disagree on the proper classification and appropriate name for certain plants. This fact is true of plants of the Bible as well as many others. The world is filled with plants, and science, as advanced as it has become, is not precisely perfect. More likely, it is we humans who are not perfect in our pursuit of science and absolute accuracy.

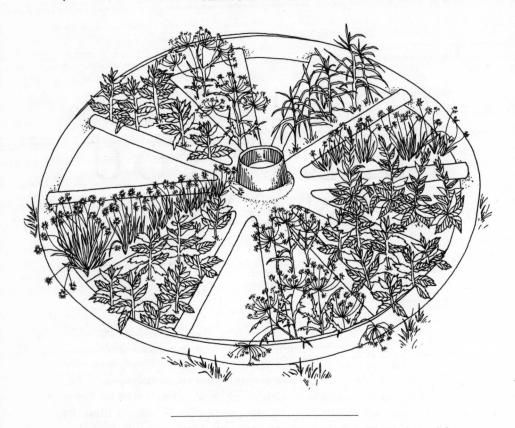

An old wagon wheel in the ground becomes a handy planting spot for some herbs of the Bible near your kitchen door.

If you research plants further through the scriptures, in old texts and more recent source books, you will find two publications especially helpful. Both these and others listed in the references of this book have proved valuable. Cruden's Concordance was written in 1737 by Alexander Cruden, who was born at Aberdeen, Scotland, in 1701. Although he originally intended to become a Presbyterian minister, his career was altered early in life by ill health. Nevertheless, his determination led him to dedicated study of the Bible and to preparation of his Concordance, which won him special honors at the universities of Oxford and Cambridge in England.

Cruden's Concordance and the updated versions covering both Old and New Testaments contain upward of 220,000 references. Based on the King James Version of the Bible, Cruden's Concordance is keyed to words of the scriptures, and provides the books, chapters, and verses in which these words were used.

You can, for example, decide to research cedar, the famed evergreen tree of great value to building and commerce in Biblical times. In addition to the scriptural reference to "cedar" from II Samuel 7:2, "I dwell in an house of cedar," you will find nineteen other references in other books of the Bible. Pursuing other terms, you can look up "cedar trees," "cedar wood," "cedars," and "cedars of Lebanon."

For extended reference to enhance and enlighten your reading of the scriptures as you garden with plants of the Bible, the Concordance can be a valuable aid.

The other extremely useful, informative, and illuminating book is *Plants of the Bible* by Harold N. and Alma L. Moldenke.

Written over a span of twelve years, the Moldenke book, published in 1952, traces hundreds of scholarly works as it guides readers to identifications of Biblical plants. This book also is based on the 1611 Authorized Version of King James and provides probable botanical names and logical alternatives in its plant-identifying process.

As you pursue your own goals, whether merely growing a few of the plants or cultivating a more extensive Biblical garden, your scope of study seems virtually endless. In truth, tens of thousands before us have read about, studied, and cultivated their own interests in Biblical plants.

For this book, it has been necessary to concentrate on those plants which meet several criteria. They must first, of course, be authentically identified plants of the Bible which grew in the Holy Land during the time in which the Bible was written, and through the time of Jesus. They must, secondly, be reasonably available today, so that you can obtain specimens, roots, bulbs, or seeds. Thirdly, they must be reasonably easy to grow, since few people have the facilities required to grow huge trees or plants requiring exotic care. Finally, the plants included in this book offer you a choice. They include not just what is believed to be the specific species, but also those which closely resemble the species, and are perhaps easier to grow successfully.

There are so many plants mentioned in the Bible that accurate identification is not always possible. In selecting the plants that should be included in this book, Biblical and botanical authorities have been consulted, and reliable texts and translations researched.

For more practical purposes, however, many of the plants that have truly been identified cannot be grown with reliability in most areas of the United States. Others that would respond to culture here are not available as seeds, roots, or rootstocks at this time.

Happily, many of the most beautiful, more striking plants of the scriptures are available to you for your outdoor home gardens or for growing indoors as houseplants and in containers on porch, balcony, or rooftop in the city. Here, for easy reference, are the names of the plants that you will find in this book, the fruits and flowers, herbs, vegetables, and trees:

Flowers

Crocus	Star-of-Bethlehem
Cyclamen	Sternbergia
Hyacinth	Tulip
Madonna Lily	Yellow Iris or
Narcissus	Yellow Flag

Vegetables of the Bible

Chicory	Lettuce
Cucumber	Melon
Dandelion	Muskmelon
Endive	Watermelon
Leek	Onion
Garlic	

Herbs of the Bible, Incenses Too

Aloe	Frankincense
Coriander	Myrrh
Dill	Sage
Hyssop	Wormwood

Fruits of the Bible

Apricots	Olives
Apples	Palms, dates
Grapes	Pomegranates
Figs	

Trees of the Bible

Acacia	Poplar
Almond	Oak
Cedar	Tamarisk
Laurel	Willow
Pine	Walnut

You'll find reliable nurseries listed at the end of the book that have seeds, bulbs, roots, or saplings of these plants available. As you read this book and the scriptures, you can plan to grow these plants of the Bible to expand your gardening horizons and enjoy the beauty and rewards these plants provide.

CHAPTER II

SEASONS
OF THE
BIBLE LANDS

Somewhere along the pathway through the centuries the idea has become prevalent that four seasons, spring, summer, fall, and winter, are a natural phenomenon. However, as any resident of northern states realizes when traveling through Florida, Southern California, or the southwestern desert areas of the United States, climates and seasons vary considerably.

Gardeners, many of whom have uprooted themselves and been transplanted by the requirements of employment or choice of retirement, also realize that horticultural zones vary considerably. Plants that thrive in warm, moist, or hot, dry southern areas are simply not suitable for northern climates. Not only are some plants extremely sensitive to slightly cold weather, many others, more tolerant, cannot survive the deep-freezing cold winters of northern gardens. Some, on the other hand, can survive but require an abundance of sun to bloom and bear fruit. Plants are conditioned to the climate and other growing conditions from which they have sprung. In this way they are much like people.

As Biblical scholars and budding botanists alike have tried to identify the plants of the Bible accurately, their lack of understanding of plant needs, especially climatic considerations, has led to errors in identification. Another failing has been that early writers from northern areas have tended to identify genus and species in terms of the similar families of plants which are native or introduced and acclimated to their areas. These are natural errors, of course, but nevertheless such determinations have often been repeated over the centuries.

To be successful in growing plants of the Bible, it is necessary to have a basic understanding of their cultural requirements. That should logically begin with an understanding of the land of the Bible, which, in reality, is a land of only two seasons. In truth, these two seasons are rather different from what you might expect. The "winter" is in reality the growing season, because it is the time of rain when plants are provided with the moisture they need to burst into bloom, to fruit and bear their harvests. It is the time when seeds sprout and rush to produce their plants, which in turn bloom, set seeds, and die, as annual plants must.

The land of the Bible, known through history geographically as Palestine, is today largely the modern state of Israel. This land is at the crossroads of three continents, Europe, Asia Minor, and Africa. It actually is more unique than this simple statement may indicate, for in that narrow strip of land much has occurred that has altered the flora of the area beyond what the climates and soils dictate.

From a map of the land of the Bible, you can readily see that it lies at the eastern shore of the Mediterranean Sea where Eurasia and Africa meet. In this restricted strip of land, woodland and desert meet and join the steppes between. Within this small area there are major variations in temperature and quantity of rainfall, two major factors in the development and support of the plant life that exists there. Countries much larger in land area, such as Great Britain or Spain, actually have fewer different plant species than exist in the land of the Bible. This chapter will explain some of the basic reasons for such a wide range of plants as they existed in Biblical times and as they grow there today.

The Holy Land was a part of the fertile crescent, one of the earliest centers of civilization. As few as fifteen thousand years ago, scientists estimate, the Holy Land was indeed a literal Garden of Eden, a land of milk and honey. Lush vegetation abounded and the fertile valleys, wooded mountains, and tropical oases provided abundant sustenance for those who lived there. As Moses led the children of Israel in their flight from bondage in Egypt through the wilderness, their goal was the Promised Land. Even then the land abounded in plenty. Times and the events that have transpired since have unfortunately greatly altered the look of the land. Fortunately, however, despite man's depredations against his neighbors and his environment, plants have been tougher and more resilient than we might suspect. That is most fortunate indeed for plants and man.

Perhaps to understand the plants of the Bible better, we should turn to the scriptures from the Old Testament and proceed through those passages to gain knowledge of the climate, the plants and how best to cultivate them today. From Deuteronomy 8:7–8 we can glimpse what the land of the Bible was: "For the Lord thy God bringeth thee into a good land, a land of brooks of water, of fountains and depths that spring out of valleys and hills; A land of wheat, and barley, and vines, and fig trees, and pomegranates; a land of oil olive, and honey."

Again in Deuteronomy 8:15–16, we can perceive how other parts of the land and its adjacent areas must have been: "Who led thee through that great and terrible wilderness, wherein were fiery serpents, and scorpions, and drought, where there was no water; who brought thee forth water out of the rock of flint; Who fed thee in the wilderness with manna, which thy fathers knew not, that he might humble thee, and that he might prove thee, to do thee good at thy latter end."

These fitting descriptions amply described the land of olden days. There were indeed the deserts which had to be crossed and deserts exist there today, inhospitable to most plants except those few especially adapted to the harsh conditions of baking sun and sparse rainfall. But even in these deserts there is a time when plant life bursts forth. Anyone who has experienced the flush of spring in the southwestern deserts of the United States can realize that with even the relatively few drops of rain in spring, seeds await that time to sprout, rush to bloom, and then scatter their seeds in turn to await the rain of the coming year.

In the land of the Bible there were, and are, areas of great fertility. Through the ages, thanks to the ravages of imprudent cultivation, much of these areas have lost their productivity. In Biblical times, however, there were indeed vast woodlands of oaks and cedars, pines and palm groves, olive trees and many others. The climate for their renewed growth remains today, and with the reforestation efforts being undertaken, and attention to wise agricultural practices, woodlands and productive orchards again rise from the land. In time, with peace, it may again be a land of milk and honey.

There are three basic climatic zones in the land of the Bible. Each has its distinctive features, but these zones are not as clearly defined as you may expect. In fact, adjacent valleys may reflect the plant life of the Mediterranean Zone while the mountains between support the vegetation of the steppe.

In places the transition from one zone of verdant growth to a less prosperous area of plant growth is gradual, while in other areas the change is abrupt, plant life changing dramatically within less than a mile from lush beauty to virtual desert conditions. There is today, in the land of the Bible, much diversity and contrast for so small an area.

The three zones of the land of the Bible are: the Mediterranean Zone, characterized by a dry, not overly hot summer and winter months of rainfall; the Irano-Turanian Steppe Zone, with long periods of dryness and a scant ten inches of rainfall annually; and the Desert Zone, hot and arid, with negligible precipitation.

As you would expect, the Mediterranean Zone affords the most favorable conditions for plant growth. Rainfall in this zone ranges from twenty to forty inches per year and occurs during the winter months. Temperatures rarely fall below freezing in this zone, but the dry period of summer serves as the dormant period for most plant life.

The rain that begins in late September to early October signals the beginning of the growing season. Scattered showers appear, developing into heavier rains during December and January and slackening during April. This warm, rainy winter season is the peak growing period for most crops, as well as the fruiting time for orchards, and blooming time for many of the flowers of the Bible lands. During the summer season, from May until September, rainfall is a rare occurrence. However, as horticulturalists explain, summer dews that collect during cool nights do provide moisture for plants. Total precipitation from these dews can be measured up to nearly ten inches during this apparently dry summer season.

Depending on the type of soil, amount of rainfall, and range of temperatures in the various parts of the Mediterranean Zone, you will find different types of vegetation. Both coniferous forests of pine and deciduous forests of oak are present, though not so common as in Biblical days. Terebinth and carob trees also are long rooted there.

The Irano-Turanian Steppe is basically a dividing zone that separates the Mediterranean Zone from the Desert Zone. Don't confuse the term steppe as applied here to the frigid steppes of arctic and near-arctic areas. The steppes of the land of the Bible are distinguished by their dryness rather than cold. Rainfall is sparse, seldom reaching ten inches annually. It occurs only during the winter season, from December to March. The remaining months are almost totally arid. This moisture factor is the key that influences plant life. Few plants can tolerate such difficult growing conditions, but annuals which can sprout, bloom, and reseed themselves in a short period do flourish. Not surprisingly, bulbous plants which can store nutrients in their enlarged roots or bulbs also are conveniently adapted for life in the Steppe Zone. Trees and bushes too have solved the difficulties of this area. They shed leaves in the dry time, much as deciduous trees in temperate climates drop their leaves in autumn as cold weather approaches.

Plants in this Steppe Zone, which actually is a relatively small area compared to the other zones, can burst into bloom at the appointed time. In winter and spring, the landscape changes dramatically from its drab, dry appearance. Bulbs send up their preformed leaves and blooms, which have been awaiting the signal of the rain. Annual flowers sprout and surge to take advantage of the short growing season, blessed by the moisture that the winter rains bring. Shrubs leaf out and bloom. In those areas favored by more ample moisture, vegetation blankets the land with foliage, fragrance, and color.

The time of this miracle of growth is brief. By late spring, the dry weather forces the land back into its grayish, brownish drabness as growth slows and stops.

The third major zone is the Desert Zone. It is, for most of the year, a land devoid of plant life. During the long, searing heat of summer, even the hardiest life is stressed into submission. No plants appear, but their roots and seeds lie secretly, awaiting their brief chance at life and beauty. This Desert Zone has climatic conditions similar to those of the Saharo-Arabian Desert. In the Holy Land, the desert is bounded on the west by Sinai, which includes most of the Negev Desert. To the south, it circles round the Mediterranean and Steppe areas and finally links into the expanses of the Arabian Desert.

In this harsh desert, rainfall is negligible, as is typical of deserts everywhere. When rains come, often less than four inches annually, they may fall in one deluge in an afternoon. What does not erode more gulleys in the landscape is soaked up by the parched land itself. Plants have little chance of survival in these desert areas. However, the annual rainfall may be more kind, falling over a period of time, which can awaken the resilient, dormant desert flora from their sleep.

There is another zone—more a subzone—which is called the Sudanese. These are the oases, or islands of vegetation, which exist amid the desert areas. These pockets of vegetation are often hundreds of miles from what would seem to be their normal, tropical habitat. Some botanists believe these oases are the vestiges of former tropical jungles rooted in the ancient past. They only persist where scant moisture is available, trapped in the soil and rock strata far below the sun-baked surface of the land.

The nature of these zones dictates to a large degree the type of plants that can grow and survive in them. Since the land of the Bible also is the crossroads of three continents, a great diversity of plants can be found there. For some species it is the eastern limit of their existence. For others, the steppeland species in particular, it is the western boundary; and for others originating in Africa, the land of the Bible is the northern limit of their range. You also may be surprised to learn that because of the abrupt changes between zones in certain areas, in a given two-mile walk from Mediterranean through Steppe to Desert Zone you can find strikingly different types of vegetation, all in a morning's stroll.

As you read the scriptures, exploring the world of plants of the Bible, this understanding of the climate and conditions of the land can perhaps explain why such different types of flora can exist nearly side by side. It also explains why some identifications of plants in the Bible are not realistically accurate.

The amount of rainfall or lack of it in these zones is the key climatic factor that determines the environment which encourages some plants but deters others. As all plants must have sun, water, and nutrients to thrive,

the amounts of these elements account for both the diversity and success
of plants that grow in each zone.

You may be surprised to realize that there is really no mention of four
seasons in the Bible. In The Song of Solomon, 2:11–13 you will read,
"For lo, the winter is past, the rain is over and gone; The flowers appear
on the earth; the time of the singing of birds is come, and the voice of the
turtle is heard in our land; The fig tree putteth forth her green figs, and
the vines with the tender grape give a good smell. Arise, my love, my fair
one, and come away."

You may search further, but even if you research through the Talmud, the Jewish Holy Book, you will find only reference to the two seasons of the Holy Land. Those are the growing period through harvest
time and the time of the resting of the land. In effect, as you will learn
from both the Talmud and the Bible, there are the days of sun and the
days of rain.

It may seem confusing, but that fact remains true for much of the
Mediterranean area. There are in some countries bordering the Mediterranean Sea longer growing periods than in others, but basically the zone
has a two-season year.

Much of the productive soil of the land of the Bible is found in the
valleys. That is especially true today after centuries of cutting of trees and
woodlands from the hillsides. Overgrazing also has led to erosion of the
slopes.

Since this is not intended as a book about the land of the Bible itself,
nor a travelogue about the land today, perhaps little else need be said
about the seasons there and the environment as it is changing even today.

However, for a clearer understanding of these plants of the Bible
which you wish to grow, and for those of you who may someday enjoy a
pilgrimage to the Holy Land, a flow of the flora in their natural cycles is
probably worthwhile. As gardeners, we should think in terms of the
growing year. In the land of the scriptures, that growing year begins in
October. It is perhaps most logical that the Jewish New Year, Rosh
Hashanah, arrives in October. That marks the beginning of the growing
year as the rains arrive to moisten the soil, stimulating the plants.

Along the streams and in the fields, on mountain slopes and elsewhere, the bulbs and corms of the well-designed bulbous plants begin to
grow. Sternbergia flowers along with early crocus and narcissus. Tulips
also raise their showy blooms, and anemones get ready for their time of
flower.

By November, rains have begun their vital task, awakening the other
dormant plants. Narcissi will be in their glory soon and crocuses dot the
countryside. The cyclamen in its dazzling beauty blooms, and farmers, as
they have for centuries, are busy in the fields.

In December the land has again turned verdant with the increasing rains. Berries and fruits are maturing and ripening as the land, in all its glory, bears witness to the beauty that comes alive again.

During January, anemones begin their colorful display, their vivid reds and purples adding to the loveliness. Irises begin their displays, and hyacinths poke their stalks higher as they begin to open.

By February more trees blossom as the iris, anemone, and hyacinth add their color and their fragrance to the sweetly scented air. This is, for many, one of the most beautiful times in the Holy Land.

In March the large iris and the tulip, arbutus, lupine, and other lovely plants combine to carpet much of the countryside. Desert flowers, providing rain has been timely and gentle, burst into bloom. With the warming winds of March comes the stimulation that sets the wild annual flowers into action. Gardens, planted and tended lovingly as the rains of winter fell, are reaching their prime.

In May, poppies and thistles, lilies and the flowers that enjoy the warmth of summer bear testimony to the miracle of life. Other earlier plants by now have begun to set their seeds as those spring blooms are ending.

By June, depending on the zone, late summer flowers appear, especially the oleanders with their displays of pink blooms, as the summer heat begins to fade the earlier flowers. Vegetable and fruit harvests continue among the groves and in family garden plots alike.

By late July and August the hot, dry winds have again regained dominance over the land. Plants wither, and the land returns to its traditional dormancy without the blessing of the winter rains to nurse plants along.

In September the final resting period of many plants nears its end. Others, those that linger latest, set their seeds, while the plants that are the first to bloom await the first sprinklings of the coming rains.

With the approach of October the winter rains will appear again, as they have for centuries and centuries before. With them the cycle is completed and the land again will begin to bloom and bear.

There may be just two seasons in the land of the Bible, but the flora there survives and thrives. It is a land of deep significance and meaning, as the scriptures have so well said.

From this land have sprung the plants of the Bible. Through the ages they have sustained the people of Israel, and Jews, Christians, and members of other faiths who have traveled through the land.

Today much has changed, yet much is as it was so many centuries ago. The plants—the fruits and vegetables, the vines and trees, the herbs and flowers—fulfill their role each year. Now, with this new understanding of the land and seasons of the Holy Land, you too can enjoy their special significance more as you grow them in your home, your garden, school, or churchyard. Each has its own meaning to share with you.

CHAPTER III

THE GOOD EARTH

There's an old saying among people who have tilled the good earth. "Take care of your soil and your land and it will take care of you." That wisdom is as true today as when it was first voiced. The basic principle underlying good gardening is improvement of the soil. The better humus you can build, the better the results you will gain. Healthy, living soil enables you to grow healthy, thriving plants.

Soil is alive. Make no mistake about that fact. Even in the poorest soils of desert terrain, tiny organisms, helpful bacteria and minute creatures, are at work underground. Every cubic foot of fertile loamy soil can have up to many millions of beneficial organisms in it. The more fertile the soil, the more microorganisms are most likely hard at work in it. Some devour organic matter. They help decompose and break down this material to improve the structure of the soil. Others work on the soil itself along with air and water to break down minerals and other elements.

Some gardens have rich, deep, highly fertile topsoil. Others have only a thin layer of good soil, underlaid with rocky or clay layers. Occasionally, in dry desert areas, there is little soil at all. No matter what you have, you can improve it. Some plants require fertile soil to produce abundantly. Others, especially some of the plants of the Bible, have survived over thousands of years because they have adapted to poorer soil and adverse growing conditions. Even those will prosper when you pay attention to good soil basics for their optimum growing conditions.

All soils have certain factors in common. Soil consists of organic matter in various degrees in different types of soils. It also contains water, air, and minerals. The proportions vary, but these major constituents remain essentially the same. You may have sandy soil, or clay, or rich-smelling loam. If you are to be successful in gardening, especially with vegetables and fruit crops, it is important to understand your soil.

Too often, in the land of the Bible and elsewhere around our globe, man has not paid sufficient attention to the soil. By removing trees and other large plants, continual planting to one crop only, and neglecting to replace soil nutrients taken out of the land by plants, man has often caused serious depletion of his natural resources. The proof of these errors is ample, including the dust bowls created in our own Midwest in this century. Even in Biblical times, the scriptures taught that the land needed a time of rest.

In Exodus 23:10–11 you can read sage advice from the scriptures: "And six years thou shalt sow thy land, and shalt gather in the fruits thereof: But the seventh year thou shalt let it rest and lie still; that the poor of thy people may eat: and what they leave the beasts of the field shall eat. In like manner thou shalt deal with thy vineyard, and with thy oliveyard."

In Leviticus 25:3–5 the admonition to allow the land a time of rest is made more pointedly: "Six years thou shalt sow thy field, and six years thou shalt prune thy vineyard, and gather in the fruit thereof; But in the seventh year shall be a sabbath of rest unto the land, a sabbath for the Lord: thou shalt neither sow thy field, nor prune thy vineyard. That which groweth of its own accord of thy harvest thou shalt not reap, neither gather the grapes of thy vine undressed: for it is a year of rest unto the land."

As you look forward to growing your own Biblical plant garden, look first to the good earth and how to improve it.

All soils can be improved with proper treatment. Your objective should be to aim for balance, in structure, texture, and porosity. When you pick up a handful of garden soil in spring and it crumbles freely in your hand, you are approaching the ideal. There are, of course, other factors, including nutrient levels in the soil.

However, the balance in the soil, the consistency of the growing medium, that layer in which plants root and obtain their nourishment, is of underlying importance. The closer you have or can build soil to a granular feel with clusters of soil that easily shake apart, the better your garden can grow. The better the good earth, the less chance of erosion and, even more important, the better your plants can flourish.

Texture of the soil refers to the majority of the particles making up the soil. They range from microscopic clay particles to the small gravel within the soil. Structure is determined by the way in which these particles are grouped into granular portions. A good soil structure is necessary so that plant roots, air, and water can penetrate to lower levels. Loamy soils, for example, usually have a crumbly structure, but clay, especially when wet, forms clods which resist root penetration.

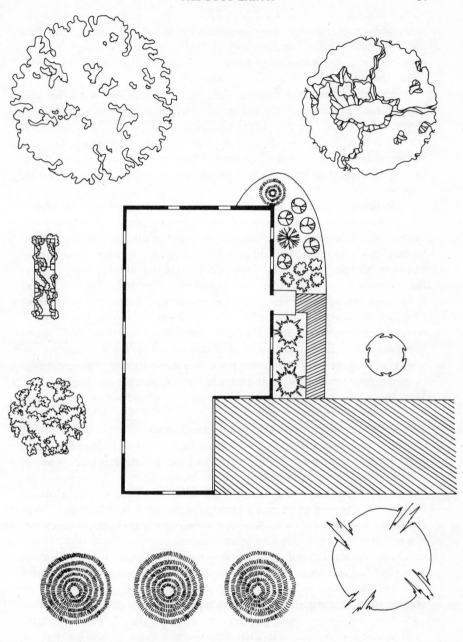

This basic landscape plan indicates how easily you can add plants of the Bible to your home grounds. The larger trees represent oak, willow, poplar, and cedar as a privacy screen. Pine and apricot also can be positioned in the front area with a sweeping willow. Along the walk, spring-flowering bulbs will add their beauty every year. Grapes on an arbor in your back yard can yield abundantly for years to come.

Productive soils actually contain up to 50 percent pore space. This porosity means that water, air, and nutrients can move through the soil. As plant roots also move more freely, they can obtain the necessary elements the plants need for lush, productive growth.

With these thoughts in mind, here are some effective ways that you can improve the soil for the plants of your Biblical gardens. There is another wise old saying that bears repeating: "No soil can have too much organic matter." That is also as true today as when first stated. Organic matter adds new life to old, tired soils. You can use organic matter easily in your soil improvement efforts. Compost and peat moss are two valuable aids.

Compost is a key. It is nothing more than decayed organic matter which breaks down from leaves, vegetation, and plant tissues. All organic matter decays in time. As it decomposes, it forms into minute particles of humus which contain some nutrients. Most important, this organic matter can open up heavy clay soils. That promotes better drainage and encourages movement of air, water, and nutrients in the soil. Organic matter from composted humus also improves sandy soil. Instead of permitting rapid loss of water, sandy soil with organic matter incorporated into it retains moisture better and longer, so that plant roots don't dry out or lose valuable fertilizer nutrients that would otherwise leach away.

Peat moss is similar to humus, since it is formed by the decomposition and compression of various types of mosses, notably sphagnum moss. However, it lacks nutrients. This material, which is widely available in garden centers, hardware and chain stores, also serves to open up clay soils and reduce moisture loss in overly sandy soils.

You can obtain peat or compost in bags and bales. However, it is easy to make your own compost. Most of the natural ingredients are at your fingertips around your home grounds. Leaves, grass clippings, vegetable scraps, leafy prunings from shrubs, all these materials can be placed in layers in a compost pile. The natural soil bacteria, aerobic and anaerobic types, work diligently to decompose dead plant material. As they do, it rots down into rich, soil-improving humus.

You can actually convert organic matter into humus in several weeks, using the Indore composting method. To do this, you simply pile layers of weeds, grass clippings, old leaves in a pile or bin. Keep it moist by weekly sprinkling if rain doesn't do that job for you.

To speed up the decomposition process, turn the pile every few days with spading fork or shovel. That simple action provides the air that the fast-acting aerobic bacteria need to rapidly decompose the organic materials. You can add manure to the pile too, which provides natural fertilizer ingredients, or scatter a few cups of 5-10-5 garden fertilizer on it as you turn the pile, to increase the nutritive values of the humus.

Another old-time saying has as much validity today as it did when pioneers began cultivating early America. It is simply, "Feed the soil, so it can feed the plants, so they in turn will feed you well."

Plants must obtain their needed nutrients from the soil in which they grow. As they grow, flower, and yield their vegetables or fruits, plants take up nutrients from the soil. You must replace these nutrients as well as improve the growing conditions of the soil if you are to have a productive garden yielding abundant harvests.

Although it is true that some plants can survive without special attention, it is equally true that tender loving care, from the ground up, will enable you to produce more glorious flowers, tastier, higher yields of vegetables, and more abundant fruit crops. In my *Inflation Fighter's Victory Garden* and my *Landscapes You Can Eat*, about fruits, trees, vines, and berry bushes, you can find many more details about soil improvement. For this book, only a basic review of good earth basics is included.

However, you will find, with each of the plants of the Bible, specific recommendations for them. One other topic is important at this point, that is: mulch. This too is a valuable ally to every gardener, since it serves you well in several ways.

Mulch is any material, primarily organic, that you apply on the soil around your plants. You can use old leaves, grass clippings, straw, peat moss, and even composted humus. The primary purpose of mulch is to smother sprouting weeds to save weeding work. By smothering weeds, you also stop them from robbing valuable moisture and nutrients from the plants in your garden. In addition, mulch helps retain moisture in the soil. It retards evaporation, thereby assuring your plants of their needed water. By using dark mulches, such as peat moss and old leaves, soil absorbs heat faster in spring, giving your plants a better spring start. The final advantage is the small amounts of nutrients which are released into the soil around your plants as the mulch slowly decays. In effect, mulch acts much like a compost pile around your trees and along your vegetable and flower rows.

As you consider the value of soil improvement to provide the best growing environment for your plants of the Bible, it pays to understand the basics of plant nutrition, too. In olden days, the use of animal manure was the only way farmers and gardeners had available to return life-supporting nutrients to the land. Today, you can use dehydrated manures and compost with other organic ingredients if you prefer to garden organically. You can also choose from a wide range of manufactured fertilizers to boost the productivity level of your soil. Fact is, you can use fertilizers that have been especially formulated to promote flower bloom or vegetable yield as well as provide the proper, balanced diet for your houseplants.

All plants need a balanced diet, just as you do yourself. The basic ingredients in fertilizer are nitrogen, phosphorus, and potash. Certain plants need different amounts of these essential nutrients, and trace elements as well. Today, manufactured fertilizers provide a range of formulations which enable you to nourish your plants for their fullest performance, be they trees, flowers, fruits, or vegetables.

All fertilizers sold have the ingredients listed in numerical order: nitrogen, phosphorus, and potash. A typical 5-10-5 garden fertilizer contains 5 percent nitrogen, 10 percent phosphorus, and 5 percent potash in the bag, or container, if it is a liquid formulation. The remainder is the carrier to insure proper mixing and distribution.

N is for nitrogen. This is the key element that promotes vegetative growth. It builds the leaves, stalks, and stems, and is vital for green-leaf tissue growth. It fosters the development of proteins, cell growth-builders in your plants. Without this essential element in the soil, you will see yellowed foliage and stunted plants. Too much nitrogen can cause problems too. Oversupply encourages excess leaf and stem growth at the expense of flower and fruit formation.

P in the formula, the second number of the three, stands for phosphorus. It is vital for strong, prolific flower development. It also promotes good fruit set and seed production and is required for proper development of plant sugars. If you like sweet-tasting vegetables and fruit, you must be certain that your plants have an adequate supply of phosphorus in the soil in which they grow.

Lack of phosphorus is readily apparent. Plants appear stunted, with a yellow look. There most likely also will be a distinctive purplish color around the edges of the leaves and between leaf veins. Lack of this ingredient also means retarded root development. Leaves may fall and plants fail to flower properly. You can correct a lack of phosphorus by applying superphosphate around your plants and raking it into the top layer of soil.

K is the final key ingredient and stands for potash or potassium. This important element promotes strong, healthy roots. It also aids in seed production. More importantly, it quickens the maturity of crops, and some scientists believe it helps increase a plant's resistance to disease.

Potassium deficiency is marked by yellowish mottling and, in severe cases, foliage loss. Roots don't develop, fruit set is poor, and plants become unthrifty. In this era of energy conservation, many people are installing wood-burning stoves and using fireplaces more frequently. Save the wood ashes. They are a fine source of potash. You can sprinkle the ashes around plants and even mix small amounts into the soil as you prepare potting mixtures for indoor plants.

Most fertilizer manufacturers offer a variety of formulations: some designed for vegetables, others for flowers, and even milder mixtures for indoor plant application.

Plants require other nutrients too, but in much smaller amounts. Luckily most soils have sufficient quantities of calcium, sulfur, and magnesium. Calcium builds plant cells and aids root growth. Sulfur is needed to help plants develop proteins. Magnesium contributes to chlorophyll, the green color in plants which enables them to turn nutrients, air, and water, with the energy from the sun, into growing activity.

Fortunately, most of the information you need to know to feed the soil, and your plants in turn, is included in the fertilizer bags or containers. Read the directions for use carefully and heed them. There is, unfortunately, a natural tendency among many of us to think that if a little is good, more will be better. That isn't so. Only apply the amounts of fertilizer as directed on the label. Otherwise, you can overburden your plants, which, like overeating by people, isn't wise. It's also wasteful.

When you plan your gardens, think of your soil as a bank. You can only take out in proportion to what you have put in. Fertilizer is actually an investment in your soil bank. You must replace the nutrients that your plants take out.

In the Biblical admonitions to let the land rest, we find the first teachings of crop rotation. If many farmers had read the scriptures more closely, they might not have planted their land to one crop year after year after year, depleting soil nutrients so badly. Constant cropping of cotton in our southern states is a prime example of this lack of respect for the land, or, perhaps, more kindly, lack of understanding of soil needs.

Improving your soil, its structure and its texture, as well as its fertility, may take time. That is time well spent. To produce the blooming beauty, tasteful vegetables, and luscious fruit from plants of the Bible, you must pay attention to their growing needs.

Even if you don't have much backyard space, or even any outdoor garden plot at all, new developments now make it possible for you to enjoy the pleasures and rewards of Biblical plants. Whether you only have an apartment or a balcony, porch or patio, container gardening is an innovative idea that may suit your needs. You can grow not only flowers, vegetables, and herbs in pots, planters, and windowboxes, but even fruit trees in large tub containers.

As the Holy Land has been made to bloom and bear fruitfully again in this modern era with carefully tailored farming techniques, you can adopt these cultural techniques also to make your own environment come alive with plants.

All plants need adequate amounts of sun, water, and soil with nutrients for proper growth. That is true enough, except that soil is not necessarily a basic requirement. What is needed is support for the plants plus the nutrients to nourish the plants. Today, you don't have to own a piece of the planet earth to grow vegetables, flowers, or fruit. Even when your share is a modest balcony on the thirty-fourth floor, you can enjoy gardening with plants of the Bible. Soilless soil has arrived. With it, and the other essentials of sun, water, and nutrients, you can indeed grow productively.

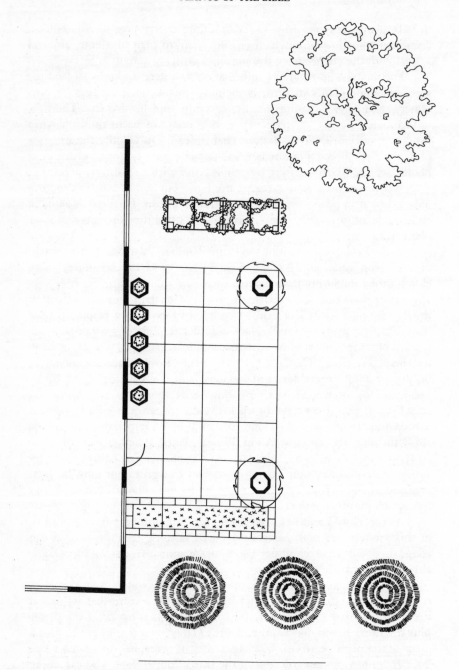

Try a patio planting of Biblical plants yourself. In this landscape design, vegetables are suggested for tubs next to the house. Herbs can be grown successfully in a raised brick bed or planter near the patio door. In larger tubs, dwarf fruit trees perform satisfactorily. You can add a grape arbor, a large willow or oak, and cedars or poplars for a decorative windbreak as well.

Wherever you live, whatever the climate, you can enjoy the plants you wish. Here are the primary considerations, with a helpful, new twist.

First step in growing plants in good garden ground or in containers is picking the right location. Your plants will need at least six hours of sun each day, some a few hours more. Water is essential since nutrients must be dissolved in water before they can be picked up by the tiny feeding root hairs on the plant roots. Outdoors, you should pick a sunny, well-drained location with at least six to eight hours of direct sun daily. Southern exposures are best. After that, eastern, western, and finally northern exposures. Even without proper sun reaching into your apartment or to your balcony, however, new developments in artificial lights make good growing possible. Both newly developed fluorescent light tubes and other conventional-style light bulbs with balanced rays closely duplicate the sun's life-giving energy. These bulbs are now widely available in garden centers nationwide.

The major new development during the past decade which makes container gardening both possible and nicely productive is the introduction of soilless soil mixes.

If you have the desire but not the plot of ground in which to garden, container gardening may be your solution. Miniature Biblical plant gardens in containers can be enjoyed on balconies or indoors. First requirement is the container, which you can buy or build. It should be, for outdoor use, of weather-resistant material. Redwood is good, but cedar is excellent. Not only do both resist decay, but cedar containers offer that extra dimension of Biblical significance. If you plan to grow large plants or small trees, be sure the container will allow sufficient room for the growing roots and is on wheels or casters so it can be moved easily.

Giant clay pots or decorative ceramic or plastic containers work well, providing you have provisions for drainage of excess water. Any soil, even heavy outdoor garden soil, that retains too much water too long can harm plants. Plant roots must breathe. If roots are surrounded by excess water, they can drown and rot. That is death on plants.

Many containers available today have built-in drainage or large saucers which catch the excess water from rain or your watering efforts. A gravel layer in them has decided advantages for preventing root rot.

The breakthrough that enables you to grow Biblical and other plants successfully in containers, wherever you live, was originally developed for the horticultural and nursery industry. This lightweight material, called soilless soil and sold as Terra-Lite mixtures, is a combination of sphagnum peat moss and vermiculite, two natural ingredients. Both materials are surprisingly light, weighing far less than conventional soil or humus. The combination mix drains well, while retaining sufficient moisture for plant needs.

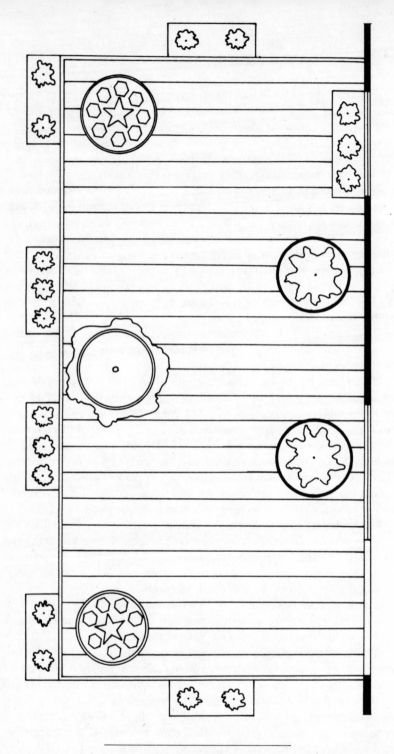

For an apartment balcony with restricted growing room, you can use windowboxes, tubs, pots and planters to grow flowers, herbs, and even small trees, selecting those species that relate to the plants of the scriptures. Be certain that all containers, including the windowboxes attached, as shown, to the railing of the balcony, have proper provision for drainage of excess water.

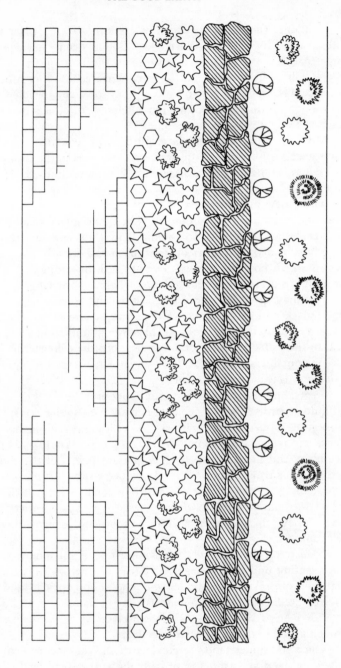

This landscape design can guide your efforts as you plant a combination flower and herb border along a path and wall. The crocuses and hyacinths can be grouped closest to the walk, with clusters of narcissi and tulips behind them as shown. Atop the wall, you can create a garden of attractive and aromatic herbs, decorative and easily accessible for your culinary use. When the spring-flowering bulbs complete their blooming period, you can use the bed for annual flowering plants.

Container gardening does require attention to several factors beyond those required for outdoor garden spots. Soil in a container will get much hotter than soil in the ground because of the buildup of radiant heat. Reflective heat from doors and building surfaces also increases the temperature. The new soilless soil mixes help solve this problem. Because they retain moisture well, but not too much moisture, heat passes through these lightweight growing mixes quickly.

More frequent watering is required with container gardens, too. That's necessary because of the greater evaporation in the summer heat of cities where temperatures rise high on balconies, rooftops, and in window planters.

These new mixtures have other distinct advantages. Since they are sterile, there are no weed problems. They do contain small amounts of fertilizer, which is included so you may sow seeds and have the seedlings properly provided with nutrients for their initial growth. After the plants have sprouted, however, it's up to you to add the additional liquid fertilizer solutions every few weeks to keep your plants thriving.

You may agree that flowers will do nicely in containers, thank you. True indeed and probably better than you believe in these soilless new mixes. You can grow poppies and chrysanthemums from seed, as well as thrill to the radiant beauty of anemones, crocuses, hyacinths, and even irises in containers. Pot herbs, from coriander to sage and sorrel, do well in such minigarden environments.

Vegetables, you may think, only grow properly in their usual place, an outdoor garden. No longer is that true. Vegetables are sprouting all across America in container gardens and yielding surprising harvests, too. Leeks, onions, and endive all thrive in containers. Would you believe that cucumbers and melons, two other important plants of the Bible, thrive there too? It's true. New varieties developed from old-time genetic pools of these Biblical vegetables are designed for small-space growing. You can realize substantial harvests from these new bush varieties, which, although new, do trace their ancestorial roots to the plants mentioned in the scriptures.

Fruit trees, too, not only survive but thrive in containers. Provided your planting unit is large enough, 2 to 3 feet across and at least 2 feet deep, you can savor the succulence of apples and apricots and sit in the shade of a cedar or fig tree on your balcony or patio.

"God helps those who help themselves" is a phrase passed down through generations. It is just as true today.

Once you understand the good earth and its modern counterpart of soilless soil, and are willing to provide the nutrients to feed your plants, you can make your home come alive with blooms of great beauty and food for more than thought. You can grow plants of the Bible anywhere you care and dare with tips, guidance, and advice in this book.

CHAPTER IV

THE CHANGING LAND

Plants of the Bible, as all plants must, rely on the good earth to give them support and the nutrients they need to thrive. Through the ages, since the Bible was written down from the poems, sermons, songs and ballads by which it was originally passed along to new generations, the soil of the Holy Land has sustained the plants as they in turn sustained the people.

Reading the Bible, we find many references to the beauty of the land, its ability to produce abundantly, both food for the body and plants that brighten the heart. If you were to visit the Holy Land today, you might be surprised to find it in many ways unlike what you have pictured from reading the scriptures. Search though you may, you will not find the apples of the Garden of Eden, the famed fruit of the tree of knowledge. For aeons the apple has been portrayed in countless paintings as the forbidden fruit, but so far as we can determine today, the apple tree was never a native plant of the Holy Land. Apples prefer cool temperatures for their best growth and cannot tolerate constantly arid conditions. In fact, considering the heat and lack of moisture in much of the Holy Land, it is most likely that the apple of the scriptures was really an apricot, quince, or pomegranate.

You can search also for the Madonna Lily, but that too is not commonly found, although wild lilies do grow there. The persistence of the lily as a plant of the Bible more likely stems from a papal edict dating back to 1618, which directed artists to include a white lily in paintings of the Annunciation.

Other plants that did grow in the land of the Bible, especially large groves of stately trees that often formed dense woodlands on the mountain slopes, are no longer there either. It is necessary to trace some of the history of the land in order to understand what drastic changes have occurred and what factors have caused the changes. The good earth that supported such abundant plant life in the days of the Bible remains, but it too has been altered through the centuries.

As the first people arrived in the land of milk and honey, they came to stay, settling down to a sedentary way of life, to till the land, plant and reap, and tend their flocks and herds. True, there were and are still nomads, tenders of flocks that wander with their charges. However, even in the earliest days, the clearing of the land began.

Trees were cut and fields prepared for cultivation, much as American pioneers cleared their land as they opened the West to settlement. As people began to plant their wheat and barley, their vines and olive orchards, natural vegetation was gradually eliminated. Homes and temples had to be built. Consequently, trees were cut. Woodlands of oak and pine, that once stood in majesty, disappeared. The stately cedars, the fabled Cedars of Lebanon, fell to the march of civilization.

Clearing the land was a necessity if people were to grow the crops that would sustain them and their children and their children's children. In fact, the scriptures urge that the land be cleared and made to bear. In Joshua, 17:15, you can read these directives: "And Joshua answered them, If thou be a great people, then get thee up to the wood country, and cut down for thyself there in the land of Perizzites and of the giants, if mount Ephraim be too narrow for thee." And, again, in verse 18: "But the mountain shall be thine; for it is a wood, and thou shalt cut it down: and the outgoings of it shall be thine . . ."

The cedars which abounded in the Holy Land were prized for construction, as cedar is today. This noble wood, like cyprus and redwood, has a natural tendency to resist decay, which makes it a desirable wood for building. Because of this property, cedar was valuable also for trade and was harvested continually through the years.

In The Song of Solomon 1:17, you can read: "The beams of our house are cedar, and our rafters of fir." And in II Chronicles 9:27, another reference to these trees: "And the king made silver in Jerusalem as stones, and cedar trees made he as the sycomore trees that are in the low plains in abundance."

In I Kings 9:11, you will find: "(Now Hiram the king of Tyre had furnished Solomon with cedar trees and fir trees, and with gold, according to all his desire,) that then king Solomon gave Hiram twenty cities in the land of Galilee." This passage and others seem to emphasize the role of cedarwood in commerce.

Cedars figure again in home building in Jeremiah 22:14–15: "That saith, I will build me a wide house and large chambers, and cutteth him out windows; it is cieled with cedar, and painted with vermilion. Shalt thou reign, because thou closest thyself in cedar? . . ."

Furniture, and other household items also were constructed of the durable cedar, as mentioned in Ezekiel 27:24: "These were thy merchants in all sorts of things, in blue clothes, and broidered work, and in chests of rich apparel, bound with cords, and made of cedar, among thy merchandise."

In Psalms 92:12, the respect with which the people viewed their cedars is apparent: "The righteous shall flourish like the palm tree: he shall grow like a cedar in Lebanon."

In II Samuel 7:2, you will find: "See now, I dwell in an house of cedar," and in I Kings 4:33: "And he spake of trees, from the cedar tree that is in Lebanon even unto the hyssop that springeth out of the wall . . ."

Throughout I Kings you will find references to houses covered with beams and boards of cedar, and walls of cedar; and cedar beams, boards, and wood carved with flowers and used for pillars.

Cedars are, in fact, one of the most widely noted trees throughout the Bible—in Kings, Chronicles, Psalms, Isaiah, Ezekiel, and The Song of Solomon. They must have been splendidly numerous.

But today the Cedars of Lebanon stand only in one lonely grove, towering witness to the ravages of man and time.

Clearing the land was one cause for the loss of the trees that graced the land of the Bible. Building, as the tribes multiplied and commerce continued, was another cause.

War was still another reason for the decline of the forests. During the times of war, armies that marched through the land often put the torch to fields and forests, denying their enemies the food and shelter that grew in their path. Often, the unprotected who tilled the soil were also put to death as avenging armies streamed back and forth across this narrow strip of land.

With destruction of the crops, fields were often abandoned as people fled before the advancing armies. Constant cycles of war pillaged the land. From the earliest times through the Crusades, battles scarred the land. During World War I, the Allied armies found the remaining forests useful again. Trees were cut to build railroad lines, and more timber was harvested to fuel the steam locomotives that rode the rails. Through the ages, in the Holy Land and across the continents, war has accounted for much destruction of the natural vegetation from smaller plants and cultivated fields to entire forests.

Despite these periodic depredations, there were times of peace when the land was again tilled and cultivated. But, as has happened many times in other areas, man has not always paid proper respect and attention to the environment. As in our own South, where cotton was grown year after year, depleting the soil, similar situations occurred through the years in the lands of the Bible.

Some types of livestock can be more harmful to vegetation than others. Overgrazing was yet another cause for changes in the ecology of the land of the Bible. Sheep and goats have a natural habit of pulling up tender grasses and plants by the roots as they graze, as well as shearing off plants close to the ground.

When more animals are grazed in an area than the plant life can support, the result is destruction of the natural vegetation.

In the Bible, we can read of this shortage of good grazing land. In Genesis 13:5–7, the conflict between Abraham and Lot is well depicted: "And Lot also, which went with Abram, had flocks, and herds, and tents. And the land was not able to bear them, that they might dwell together: for their substance was great, so that they could not dwell together. And there was a strife between the herdmen of Abram's cattle and the herdmen of Lot's cattle . . ." Flocks tended by Biblical nomads, like those of today's Bedouin, continued to overgraze areas until they were no longer productive.

In time, rains and winds eroded the topsoil, which was left unprotected by vegetation. Slopes and mountains, denuded of the trees, also became eroded. Rains carried soil that had been held in place by tree roots down to the valleys, further upsetting the balance of nature.

Today, in this era of growing interest and concern for our environment, we are becoming more aware of the delicate balances of ecology. And today in the Holy Land, reforestation projects and wiser agricultural development with irrigation systems are changing the face of the earth once again. Fortunately there is a new appreciation for soil and plants which is bringing vegetation back to otherwise barren land.

Through the centuries, other plants of the Bible were also threatened with extinction. Pilgrims traveling to worship in the Holy Land, as travelers will today, picked and uprooted plants which they wanted to take back to their native lands. Some of the rarer plants, deprived of the opportunity to produce seeds for the following year's growth, were reduced in numbers. Even today, in America, we have not become fully aware of the need to protect the dwindling numbers of rare plants. Even with an endangered species law, some plants, like some animals, still dwindle in numbers toward the point of extinction.

In the Holy Land, all these factors have had thousands of years to influence the fate of plants. Fortunately, plants have astounding persistence. As the hyssop, which forces its roots into the tiny slits in ancient rock walls, and as trees that can often find a roothold on a precarious ledge of a mountain slope, plants seem at times exceedingly tenacious.

The soil of the Holy Land, in which the many plants grow, varies from area to area. Much of the land is sandy. Even when adequate rains do fall, sandy soil has an obvious tendency to soak up the water, letting it seep deeply into the loose earth, where many plant roots cannot reach it.

The soil and silt that erodes from denuded slopes also creates problems, mixing with more fertile valley soils and, at times, burying the good earth under less desirable soil, sand, and silt.

When you consider that these processes, both natural and man-caused, have been occurring for thousands of years, it is easier to understand the stress that has been placed on the flora of the Holy Land. Perhaps this process can better be put into perspective with some thoughts and figures from *Future Shock*.

If the last fifty thousand years of man's existence were divided into lifetimes of approximately sixty-two years each, there would have been eight hundred such lifetimes. Of these, only during the last seventy lifetimes has it been possible to communicate effectively from one lifetime to another in writing. Only during the last six lifetimes have masses of people seen the printed word. Only during the last few have people seemingly begun to enjoy the abundance of life that our modern era has made possible. Now that we can communicate so well, in printed word, and electronic media too, and have such potential at our command, we also have an obligation.

Today, as we attempt to study and understand the scriptures, and learn how to cultivate the plants around, especially the plants of the Bible, we must realize another fact of life. We are truly stewards of the land for a short period of time. We are responsible for understanding, protecting, and preserving the land and the plants that grow from it. Perhaps as stewards we should pay more attention to preserving our land and the land around us, as well as the plants that depend on the land for sustenance. By holding greater reverence for all life, the plants included, we may, as stewards, do our small share in improving the fragile growing world around us. We must, if future generations are to enjoy the glory and the beauty of a wide diversity of plants which we enjoy today.

As we plan and plant, till and tend our own home gardens, there is ample opportunity to look beyond our own private horizons. As community gardens and city beautification projects have sprouted, so can new beauty in and around our churches and synagogues. You may find common bonds with other gardening friends to launch growing projects through church groups and societies.

In one area of New Jersey, garden club members sponsors a "deck the hills with daffodils" program. Each spring, the hills of Somerset County comes alive with tens of thousands of daffodils. These are closely related, of course, to the narcissi of the Holy Land. Elsewhere, church members have planted churchyards and the homes of their pastors with herbs and flowers, carefully researching the plants to provide miniature replicas of gardens that existed in Bible days, so many centuries ago.

In other areas, in Iowa, Wisconsin, Florida, Arizona, plant enthusiasts have been using seeds and bulbs of Biblical plants in their Sunday

School classes for several years. As youngsters learn to read the scriptures, they also have learned to relate to the plants and the natural history of the Holy Land. In this small way, they too gain greater appreciation for the living world around them. And the plants they tend so lovingly serve well as gifts for shut-in members of the church, as well as for parents during holidays.

Perhaps, when you survey your growing world, you and your friends will find new horizons beyond your own dooryard where gardens can bloom and bear. Plants of the Bible can provide not only beauty to the eye but also deeper enrichment and understanding of the scriptures, too. From these gardens, our faith and heritage can grow more deeply in the years to come.

CHAPTER V

BIBLICAL GARDENS YOU CAN VISIT

You can inhale the sweet fragrance of Biblical gardens as you enjoy the beauty of their flowers, trees, and other plants in several places in the United States. Undoubtedly, many small plantings exist not only on the grounds of religious orders, but also in many churchyards across America. However, the devoted efforts of plant enthusiasts have led to the creation of more extensive Biblical plant gardens in several areas which you can visit. As you do, you'll not only savor the pleasure of these gardens, but will be able to transplant some of the ideas you find there to your own home grounds for your personal gardening pleasure. In fact, you may be inspired to organize efforts with other friends and members of your church, synagogue, or community to plan and plant Biblical gardens in your own home town or city.

The Biblical Garden of the Cathedral Church of St. John the Divine is one of the most extensive and inspiring of such gardens. Located on the thirteen-acre complex in which this Cathedral stands on Morningside Heights in New York City, it contains hundreds of plants representing species that existed in the Holy Land two thousand years ago. Thousands of visitors each year come to visit this garden, marveling at its beauty and learning more about these plants and their place in the scriptures.

At the time of Christ and even long before in the Holy Land, most gardens were primarily functional, providing the food and fruit for sustenance of the people who lived there. Gardens also were grown for their beauty, and served as sanctuaries where people could find refuge, tranquillity, and peace. These gardens contained trees, flowers, shrubs, and

fruits, and many other species of plant life. Some trees were intended for
food, of course, but others were planted for shade and beauty. Most plants
served a dual purpose, from the herbs that had their medicinal values to
the fruits which offered lovely displays of blooms before producing their
fruitful harvest.

The idea of a Biblical garden at the Cathedral Church of St. John the
Divine emerged as the inspiration of Sarah Larkin Loening. Searching
through the Bible and assorted texts about Biblical plants, Mrs. Loening
realized that such a garden not only would provide inspiration to the thousands of pilgrims who visited the Cathedral each year, but also would
serve as a place for meditation and a living garden of study for children.
Through her leadership and with the assistance of C. Powers Taylor of
the Rosedale Nurseries of Hawthorne, New York, this garden was lovingly planned and planted. Today, with its apricot and quince trees, willows, oaks and cedars, nut trees and grapevines, it has indeed fulfilled its
intended purpose. In addition to the trees, herbs and flowers abound and
are clearly identified so that all who visit can better appreciate the significance of the Biblical plants there.

This leafy guide to the Bible was begun in 1973 and today its
hundred different species provide inspiration to the many thousands of
visitors who are attracted to it annually. Nestled against the south side of
the towering cathedral walls, the quarter-acre plot inspires an ecumenical
following. At times youth groups representing Jewish and Catholic as well
as Protestant schools and organizations arrive to explore the florae of the
Bible. Among fig trees and carobs, pomegranates and cedars are smaller
plants.

The Judas tree or *Cercis siliquastrum*, the tree from which Judas
hanged himself, according to tradition, is represented by the more hardy
Cercis canadensis, since the actual species is not hardy in this zone.

Try as you might, however, you will find no apple tree. The best
botanical opinion over the years is that the apricot is actually the "apple"
of the Garden of Eden. Apple trees, as we know them today, are native to
northern areas and do not grow in the typical Mediterranean-type climate
of the Holy Land. However, you will find a re-created garden of papyrus
in a tiny pond, and a clump of wormwood, cited in Deuteronomy 29:18.
In early spring, bright stars-of-Bethlehem burst forth sprouting from their
tiny bulbs. In II Kings 6:25, "the fourth part of a cab of dove's dung" is
cited as reference for this plant, since its white flowers appear to sparkle
on the hillsides of Palestine as the white droppings of doves might.

Strolling through this garden, you may be surprised to discover
familiar plants too. Sage, dill, leeks, endive, onions, lentils, and sorrel are
grown here. Among the flowers, anemones, narcissi, tulips, lilies, and
irises bloom in profusion in season, all clearly identified with the appropriate scriptural references.

Well rooted now, the garden blooms in season, gracing the grounds of the great Cathedral. The church itself is the largest Gothic cathedral in the world, and work still continues toward its completion. When finished, the Cathedral will hold more than ten thousand worshippers. As a church, it is second in size only to St. Peter's in Rome, which is actually a basilica rather than a cathedral. St. John's has been described as "the Word in Stone," as a sacred book written in stained glass and great arches. To this perhaps should be added the message from the living garden which now thrives there. The Cathedral was begun in 1892 and is built almost entirely of stone, in the manner of the medieval cathedrals. Of special interest are the great organ and the great rose window and the lesser rose window below it, in the West Front, as well as the other superb stained-glass windows.

In the Biblical Garden you will find several dozen different plants of the Bible. The hardy plants that can survive the rigors of winters in New York City, in horticultural zone 6, remain permanently in the garden. Many others, being less hardy, are planted in containers and are on display in the garden during the spring, summer, and fall. Still others are planted annually from seeds, growing, flowering, and contributing their special significance in the overall Biblical plantscape.

As you wander through the garden, you'll find some of the exact species of plants which learned botanists have determined to be the plants mentioned in the Bible. In some cases, however, the exact species were not available, or would not survive in the temperate climate and severe winters. In such cases, closely similar species or cultivars of the same genus have been planted. Among the plantings you'll find common names, Biblical names and references, and, finally, the botanical names. The scriptural references listed have been taken from the excellent book by Dr. Harold N. Moldenke and Alma L. Moldenke, which is one of the most reliable botanical texts about plants of the Bible. In this book also, as in the Moldenke book, scriptural references refer to the Authorized Version of King James I published first in 1611.

For those interested in participating in the growth of this garden project as it is expanded, membership in the Biblical Garden is available. Information about the garden and membership may be obtained from Mrs. Albert P. Loening, Biblical Garden, 1047 Amsterdam Avenue, New York, New York 10025.

Another excellent Biblical garden, a sanctuary of peace and inspiration open to the public, is in Coral Gables, Florida, at St. James Lutheran Church. There, the Garden of Our Lord is planted with shrubs, flowers, and trees native to the Holy Land. An advantage of this southern climate is that the plants can thrive outdoors in the warm Florida sun throughout the year. Some of the plants that grow there can only be grown indoors in northern climates, but from this garden you also can transplant many ideas back to your own home gardens, indoors or outdoors, depending on where you live.

From the first chapter of Genesis, with its poetic story of creation, to the last chapter of Revelation, plant life is significant to the natural flow of history, prophecy, and beauty. During the past few years, a surprisingly large number of Biblical plants and shrubs have been established in this unique project, the Garden of Our Lord. It is surrounded by a stone wall, which provides a suitable setting of privacy for meditation amid the diversity of Biblical plants.

Under the cool shade of the acacia groves, the Israelites were led in worship; and here you too can stand beneath their shade. As you recall the story of Noah, from Genesis, when the dove returned with a silvery olive leaf in her beak after the floods, you will find olive trees as well as fig trees in this garden. Recalling the passages from The Song of Solomon, you will find aloes, hyssop, cinnamon, and other plants growing in the subtropical climate of Coral Gables.

The almond, palm, and pomegranate were fundamental plants to the people of Biblical days in their life in Palestine. These too grow in the Garden of Our Lord, arrayed with myrtle, lilies, and other flowering plants. Cedar trees, prized of old for their power and strength, and so widely used in building in the land of the Bible, stand stately in these gardens. Willows too, on which the captive Israelites hung their harps along the rivers of Babylon, are included in the garden's groupings.

There also is a Children's Pool, where tropical fish abound, and in which an inspiring statue of Christ, hand-hewn in Italy from an eight-ton block of Carrara marble, is reflected, an arm upraised in a gesture of benediction.

In these gardens you will also find the beauty of flowers grown from seeds gathered in the hallowed spot in the Kidron Valley, the Garden of Gethsemane. The Judas tree, passion flower, crown-of-thorns, and many other plants long associated with the Christian story in song and legend also thrive in this Florida Biblical garden. It has been designed to provide a clearer picture of the everyday lives of men and women who made Bible history, as it helps us realize the significant role that plants played through all the scriptures.

The Garden of Our Lord in Coral Gables is located just east of Ponce de Leon Boulevard, four blocks south of South West Eighth Street, the Tamiami Trail, and one and a half blocks west of Douglas Road, on 110 Avenue Phoenetia. For anyone wishing to marvel at some of the most famous plants of the Bible in a natural setting, this Garden of Our Lord is a worthwhile stop on your travel agenda.

In Israel today you can find what is most likely the largest, most extensive Biblical garden of our world: Neot Kedumim, the Gardens of the Bible. Standing between the Judean foothills that stretch toward ancient Jerusalem and modern Tel Aviv, this five-hundred-acre garden has been designed to explain and illustrate the Bible rather than merely re-create the era of the Bible. In effect, it does both, and encompasses both Jewish and Christian tradition, since the roots of both religions are so deeply rooted in the Holy Land.

Neot Kedumim, sprouting more extensively each year, is the fulfillment of a dream shared by Dr. Ephraim and Hannah Hareuveni, two Russian Jewish emigrants who envisioned these gardens of both flora and fauna. Both trained botanists, the Hareuvenis dedicated themselves to research of the land and ancient literature of Israel. They conceived the idea of creating a living replica to reflect the history and traditions of the land of the Bible which would create a living bond between the past, present, and future. It thrives today, a tribute to their dedication, and the perseverance and the dedication of their son, Nogah Hareuveni.

In Neot Kedumim, visitors can wander among groves of palm, fig, and olive trees, observing the intertwining of Jewish and Christian traditions. In these extensive gardens, olive trees stand symbolically as they have since ancient times. According to universal tradition, olive branches are a symbol of peace, and they are included in the state emblem of Israel.

In Exodus 27:20, God commanded that olive oil was to be used in kindling the light of the menorah because it provided the brightest and steadiest flame. Also prominent in the gardens is the moriah plant, which is a member of the sage family. This herb, which grows virtually wild throughout Israel today as it did in Biblical times, has special significance to the Jewish people. The menorah is first mentioned in the Bible when God instructed Moses in the preparation of the Ark of the Covenant. As described in Exodus, 25:31–40, the specifications seem couched in botanical terms of branches, calyxes, cups, and petals. Ancient Jewish scholars point to a direct relationship between the menorah and the moriah or sage plant as a particular plant. The moriah may not always have seven branches, but it does have an even number growing from a central branch, and its pattern is strikingly similar to the menorah.

This marvelous garden is located at 1047 Amsterdam Avenue on Morningside Heights in New York City and is open from 8 A.M. to dusk. People who work and live in the neighborhood often bring bag lunches to the Biblical Garden, where specimens of plants mentioned in the Bible are carefully tended by volunteers who call themselves the Divine Gardeners.

Information about Neot Kedumim can be obtained from them directly at P.O. Box 1007, Lod 71 100, Israel. Their telephone is (08) 233-840 and Fax number is 972-8-245-881.

The United States representative for Neot Kedumim is Paul Steinfeld. He can be reached at Gilead Tree Farm, Steinfeld Road, Box 63, Halcott Center, N.Y. 12430, by telephone at 914-254-5031 and Fax at 914-254-4458.

In Pittsburgh, Pennsylvania, you'll find another fascinating Biblical

Garden, Rodef Shalom Biblical Botanical Garden. Located on a third acre of Rodef Shalom Temple this fine garden lets you explore the natural world of ancient Israel with more than one hundred temperate and tropical species. You can enjoy the plants amid a setting of a cascading waterfall, a desert, a bubbling stream representing the Jordan River which meanders through the garden from Lake Kineret to the Dead Sea.

During summer, a large group of hosts and hostesses are available to the thousands of visitors that arrive each year. This garden is located at Rodef Shalom Temple, 4905 Fifth Avenue, at Fifth Avenue and Devonshire Street in Pittsburgh. It is open from June through mid-September and a harvest festival is held each fall. Information is available from Rodef Shalom Biblical Botanical Garden, 412-621-6566.

When in Charleston, South Carolina, visit The Biblical Garden at Magnolia Plantation. The garden is divided into three parts. The outer perimeter is devoted to trees and shrubs of the Bible. The interior is divided between the Old and New Testaments, featuring bulbs and flower plants mentioned in each of the Testaments. The Old Testament area is arranged into twelve small beds forming a Star of David and represents the twelve tribes of Israel. In the center is a statue of David. The New Testament area is arranged in the shape of a cross, featuring a statue of the Virgin Mary, with twelve beds representing the twelve disciples.

Other Biblical Gardens worth trying to see include those at the Pleasant Valley Baptist Church in Camarillo, California, Church of the Wayfarer in Carmel, California, the extensive St. Gregory's Episcopal Church Bible Garden at 6201 East Willow Street in Long Beach, California. Others include Church of the Holy Spirit in Orleans, Massachusetts, and Beth El Bible Gardens in Providence, Rhode Island.

As I completed this book, it became obvious that many people and groups from all across America would enjoy and like to see and discover more about the plants of the Holy Land in the plants' native habitat. For that reason, I am developing special tours called Gardens of the Holy Land Pilgrimage Tours.

You may obtain additional details about these tours to the Holy Land from UNITOURS, 411 West Putnam Avenue, Greenwich, Connecticut 16830. You may also get information through travel agents or directly from me: Allan A. Swenson, Windrows Farm, RFD #1, Box 1987, Kennebunk, Maine 04043.

PART TWO

CHAPTER VI

FLOWERS OF THE BIBLE

The land of the Bible, unique as it is in climate and seasons, gave birth to an amazingly diverse array of flowers. In fact, Asia Minor and the countries bordering the Mediterranean Sea are the cradles of the classic bulbous plants so popular in Europe and across America today. From these regions have come our tulips and hyacinths, crocuses and narcissi, lilies and irises and cyclamens. In this crossroads of three continents, bulbous plants developed naturally under the conditioning of the hot dry season, which forced them into dormancy, and the brief moist period, when they must burst forth rapidly.

The term "bulbs" is a horticultural term. It includes the true bulbs, which have food storage parts of the plant surrounding and protecting the preformed tiny plant bud itself. If you cut a tulip bulb in half, you can easily see that growth bud nestled securely in the center, the makings of the complete tulip: leaves, stem, and flowers. The general term "bulb" also includes plants that rise from similar bulbous roots: corms, rhizomes, and tubers. These plants are a study in survival and adaptation.

Since many of the flowers of the Bible are plants that grow from bulbs, perhaps a closer look at this fascinating form of life will be worthwhile. A true bulb includes a central growth bud which is sheathed in graded layers of modified white leaves. These are known as scales on such plants as lilies, narcissi, and tulips. Corms are slightly different, being bulblike but solid, consisting of a fleshy base of the stem which stores up the needed nourishment that was gathered by the leaves during the growing period.

In effect, the same is true with bulbs, but that food reserve is stored in the scales that surround the tiny growth bud. A crocus grows from a corm, as do some other plants, including gladiolus.

*To force spring-blooming beauty in your home from the bulbous flowers of the scriptures,
plant bulbs in a low pot or bulb pan by early October for liveliest displays by January.*

Among bulb-type plants, some have tubers, which are thickened
underground stems with rough, leathery skin. Begonias and cyclamens
grow from tubers. Another type of "bulb" is called a rhizome. This is a
thinner, more elongated underground root than the tuber. It is formed by
the stem of the plant and is, in fact, an underground stem or rootstock
that bears buds. The iris is typical of plants that grow from rhizomes.

This book isn't intended as a botanical text. However, some of these
facts may prove worthwhile in the culture of your plants of the Bible
around your grounds. More likely, an understanding of these plants will
prove useful when you begin to enjoy the glorious blooming beauty that
comes from forcing bulbous plants to bloom at your wish and whim
indoors.

When the snow is falling and winter winds are wailing, there is nothing quite like the dazzling displays of color you can enjoy from bulbs you

Narcissi, tulips, and hyacinths join together to welcome spring with pine and willow trees in this lovely setting for a porch or patio Biblical plant garden.

force into early bloom. With proper timing, you can provide yourself and friends with grand displays of narcissi, tulips, fragrant hyacinths, and even profusions of crocuses blooming on your windowsill from late December through March.

These hardy plants that sprout from bulbs of various types have won wide favor in many parts of the world. The bulb industry that has grown in Holland is world-renowned. With help from these dedicated plant breeders, you now have an amazing choice of colors, shapes, and sizes among the tulips, narcissi, and hyacinths that you can grow to grace your outdoor plantscapes. Aside from forcing a few bulbs for winter beauty indoors, you can utilize this wide selection of bulbs in beds and borders, and in more formal landscape designs. You can use narcissi and daffodils in naturalized plantings beneath trees and shrubs, along walls and fences.

Only two flowers appear by name in the Old Testament. These are the rose and the lily. There still is much debate concerning the true botanical identity of the rose and the lily. In all likelihood neither is the flower it seems to be according to its name. On that, most Biblical scholars and botanists agree. Remembering that the Bible was not intended as a text on natural history, the plants which appear in the scriptures are primarily significant for the idea they illustrate and the lessons they illuminate.

Palestine was, as Israel is today, a land graced with many wild flowers. Anyone traveling to the Holy Land will find fields abounding in the beauty of these flowers. Prominent among the plants are those bulbous types. Unlike the crocuses, hyacinths, narcissi, and tulips that burst forth in their glory in spring in the United States, these plants await the coming of the winter rains to perform their colorful feats in the land of the Bible.

From October to December, you will find them blooming in the fields of Israel. In fact, the climate is so favorable to these bulbous plants that a horticultural industry is rapidly growing there today. It may not yet rival the millions of bulbs produced in the Netherlands, but it is substantial and growing more important every year.

As you plan your flower gardens, focusing on those that have their rooted heritage in the scriptures and the Holy Land, you can follow the growing advice in this chapter. With these tips and ideas, your home and grounds will produce abundantly, reminiscent of the gardens and the flowers that glorified God in the Holy Land.

ANEMONE

"And why take ye thought for raiment? Consider the lilies of the field, how they grow; they toil not, neither do they spin: And yet I say unto you, That even Solomon in all his glory was not arrayed like one of these." St. Matthew 6:28-29

"Consider the lilies how they grow: they toil not, they spin not; and yet I say unto you, that Solomon in all his glory was not arrayed like one of these. If then God so clothe the grass, which is today in the field, and tomorrow is cast into the oven; how much more will he clothe you, O ye of little faith?" St. Luke 12:27-28

How could the lilies of the field be other than lilies, you may well ask. The lily is probably the most famous of all plants of the Bible, even more so than the rose. It is, of course, possible that the lily mentioned in

The stately Madonna Lily traces its heritage to traditions from the scriptures. (Photo courtesy Parks Seeds)

the scriptures is a lily, since lilies are bulbous plants and some examples of these plants were and are favored in the land of the Bible. Another fact that points to the lily that we know today is its longtime association with holy days. Many paintings and carvings of the Madonna and of the Resurrection and Ascension show the Madonna Lily prominently. In fact, the major artists of the ages, Titian and Botticelli among them, have all painted the Virgin Mary with a white Madonna Lily. A papal edict was issued in 1618 which laid down stringent rules concerning the appropriate treatment of certain sacred subjects in art. In this edict, the necessity of showing the white lily in paintings of the Annunciation was emphasized. In addition, palms and lilies were to be shown as flowers suitable to be scattered about by angels.

From those days, and reflected in so many works of art, the Madonna Lily, *Lilium candidum*, has gained a strong following in its contention as the lily of the fields. Lilies through the Greek and Roman periods also were traditionally symbols of purity and grace. To this day, at Easter, lilies grace the sanctuaries of churches across America and around the world. Most of these are actually variations of the *L. candidum*, as florists and growers point out.

With what seems to be so much proof that the lily is the lily of the fields, it is difficult to believe otherwise. However, delving into the botanical facts concerning lilies of the field, there is much room for debate. The Madonna Lily is not really common in Israel. Another fact must be understood. In translations of the scriptures from the ancient languages, the same word has been used to mean the tulip, the anemone, and the iris, as well as other showy flowers, depending on who has done the translation.

If you examine the iris, it can be said to be as showy as a lily. The cyclamen also has been said to be the lily, since it grows wild today as it has for thousands of years in the Holy Land. One researcher, writing on plants and trees of scripture in 1851, acknowledges the many paintings of the lilies but firmly names *Sternbergia lutea* as the lily of the fields.

Dr. James Moffatt and Dr. Edgar J. Goodspeed also undertook translations of the Bible during their lifetimes. What makes their more modern translations different from earlier versions of the Bible is that they dug deeply into botanical and scientific terms and plant descriptions. With that in mind, it is understandable that these men might suggest other plants as they searched through early texts and scriptures. Dr. Moffatt seems firmly to believe that the rose is really the narcissus, *Narcissus tazetta*. Considering the works of Drs. Moffatt and Goodspeed, and many who also have sought the true identity of the lily, rose, and other flowers, most authorities today believe that several plants were actually meant.

Among Biblical authorities, many point to the belief that Jesus always chose the most familiar object in the daily life of the people to illustrate His teachings. This belief leads many to state that the anemone was truly the lily. It did and does grow wild in profusion throughout the land. Today *Anemone coronaria* in its various forms can be found throughout Israel, blooming on the shores of the Lake of Galilee to the plains and foothills beyond. The anemone is a bulbous plant that produces showy flowers with five to seven brightly colored blooms in scarlet, purple, rose, and yellow. With its brilliant colors, it is understandably one of the most conspicious plants in the Holy Land.

Since most authorities regard the Palestine anemone, *Anemone coronaria*, as the famous lily of the field, we are inclined to accept the majority viewpoint. However, in deference to those who hold other views, you will find in this chapter details about growing anemones, tulips, and cyclamens, as well as narcissi, hyacinths, and lilies.

The anemone or windflower is a member of the buttercup family, which contains about 120 species. These plants are perennial herbs with an underground rootstock. Some types are tuberous-rooted. The plants have radical, deeply cut leaves with elongated flower stems. The *Anemone coronaria*, also called the poppy anemone, is the most likely candidate for the lily of the fields, most authorities now agree. This tuberous-rooted plant has divided leaves and large, showy, poppy-like blooms. Flowers may be scarlet, crimson purple, blue, or white. Although plant breeders have developed others, including double-flowered types, the best for most gardens is the poppy anemone. Fortunately, these plants also can be forced, as florists often do, for bloom in midwinter. They also thrive in cool greenhouse conditions.

Anemones are easy to grow when started from tubers. They grow well as a ground cover and prefer light shade. This fits them well for use in rock gardens, under shrubs, and in naturalized masses beneath trees.

Plant anemone tubers about 2 inches deep in a cool, moist, and well-drained soil. In northern states, plant your anemones in the early spring. In southern areas, you may plant them in the fall, and cover with mulch during the first winter.

Anemones will sprout readily with adequate moisture and will thrive in somewhat dry soils. If you prefer to grow a few in pots, a standard potting mixture is satisfactory. Plant your tubers, as florists do, 2 inches deep and water regularly until they sprout.

To prepare your own soil mixture, use equal parts loam, leaf mold, and sand. Be sure that you place the top side of the tuber up, whether in pots or outdoors. It can be recognized by its somewhat fuzzy appearance. The bottom of the tuber is usually pointed.

Anemones are also excellent for cutting from your outside garden if you don't elect to grow some as potted plants. If you wish to expand your anemone culture skills, there are other related anemones available, including the double-flowering types. However, if you agree with the majority of Biblical scholars and botanists, the poppy-flowered, single anemone, A. *coronaria*, should be your first choice.

CROCUS

"Spikenard and saffron; calamus and cinnamon, with all trees of frankincense: myrrh and aloes, with all the chief spices . . ." The Song of Solomon 4:14

Fortunately, in tracing the roots of Biblical plants, it is occasionally possible to find one identification on which almost all authorities, Biblical scholars and botanists alike, agree. The crocus is most assuredly an authentic plant of the Bible.

Saffron was a valued dye in the time of the Bible, as it is today in many countries far beyond the borders of the Holy Land. Considering the fact that it requires at least four thousand stigmas and upper portions of the style of the blue-flowered Saffron Crocus to make an ounce of saffron, it is one of the most expensive products of its kind. No wonder it was so valued in the time of the Bible, even as it is today.

The Saffron Crocus, *Crocus sativus*, is native to Asia Minor and Greece, and is found also in other Mediterranean countries. After the stigmas and styles are gathered, they are dried in the sun, then pounded into small cakes. This expensive product is used primarily as a yellow dye, and also for coloring in curries and Oriental foods.

Actually, many types of crocus grow in Palestine as they did before the birth of Christ. You can find white and pink, blue and yellow as well as lilac and purple types of crocus in the fields, rocky hillsides, and along roadsides, but only the blue-flowered types produce the saffron. These must have grown in great abundance to have withstood the constant harvesting which the demand for saffron created. In ancient times, old records reveal that saffron was scattered during wedding ceremonies and mixed with wine. It has been used to make perfume and to color confections as well. Some old texts indicate that saffon was given medicinally as a stimulant and with spices and flower petals as scents to perfume rooms.

In other passages of the scriptures, in the translation of Dr. James Moffatt between 1922 and 1925, the good doctor takes pains to change some of the wording of The Song of Solomon. Where, in 6:4, it is written in the Authorized Version, "Thou art beautiful, O my love, as Tirzah,

Dozens of colorful crocus blooms greet spring each year, reminding us all of the crocus plants that grace the land of the Bible.

comely as Jerusalem . . ." Dr. Moffatt is of the view that no love song would compare a young maid to cities. Therefore, he renders the passage as "You are fair as a crocus, my dear, lovely as a lily of the valley." While it may seem more appropriate to compare a young girl to flowers rather than to cities, other scholars believe that Dr. Moffatt has taken some liberties in his translation.

However, if you feel as he did, that a crocus is a more fitting comparison, you may then wish to read further among Biblical scholars. Many believe that if The Song of Solomon was actually composed by Solomon, it was done somewhere in the hills and mountains on the slopes of Lebanon. There, botanists report, four different species of crocus do occur in profusion.

Even if you forego this identification by Dr. Moffatt, the original interpretation of saffron in reference to the product of the *Crocus sativus* is on solid ground.

Whichever way you wish to vote, you can enjoy the beauty of modern crocuses around your home and garden as the ancients no doubt did.

The Saffron Crocus may closely resemble our springtime crocus, but actually it blooms in its native habitat in late autumn. You may choose the purple-blue spring crocus for your garden or the purple fall crocus, *Crocus zonatus*, that also thrives across northern regions of America. If you happen to live in hot southern areas, you may find it difficult to have success with these plants and, in fact, other spring-flowering bulb flowers such as tulips, narcissi, and hyacinths. All these plants require a period of cold weather in which they can go dormant. Nevertheless, you can enjoy them by following the guides for forcing varieties of these bulbs indoors. That is covered in a special section following this chapter and makes an exciting indoor project for children of all ages.

Crocus bulbs are easy to plant. No garden should be without these cheerful welcomers of spring. By planting the largest, earliest flowering, you often will be greeted by their colorful blooms even before all snow has melted. Today's crocus varieties have been specially developed for more profuse blooms and showier colors than the original wild species. Truest in appearance to the scriptural plants are the blue ones, whether the species you plant are for spring bloom or for fall. Some purists prefer *Crocus zonatus*, since these are wildings and multiply rapidly without care or trouble.

Plant autumn crocuses in the spring, in sun or light shade. As perennials, they will be a permanent part of your plantscape beneath trees, under shrubs, in beds, and along pathways. Once planted, they will increase year after year to produce profusions of blooms. This species is best left to its own resources, which does save your gardening time and effort.

Spring-flowering crocuses should be planted between mid-September and early November, before the ground freezes. New varieties, related distantly to the crocus of the scriptures, offer larger, more abundant blooms. Plant bulbs 3 to 4 inches deep in clusters or groups. Once planted, they are so persistent that even a rank amateur can obtain perfect results. Spring crocus and the earlier flowering crocus species multiply year after year. For best results, it pays to obtain the larger bulbs which cost a little more but do yield larger blooms.

You can also force crocuses in bowls or glasses. Some firms offer crocus-forcing kits, including ceramic pots, with holes through which crocus plants can grow. For forcing bulbs you buy locally, use a Terra-Lite mixture of peat moss and vermiculite. Place the bulbs in the mix with the points aimed at the holes, and the other few bulbs on top of the container, just the tips emerging from the soil mix. Water thoroughly and keep moist as bulbs sprout in a cool, dark spot. When the sprouts are 2 to 3 inches tall, bring the container into your display room, but keep out of direct sun. An eastern window is best, or place the container back from a hot southern window. For weeks, you'll be radiating in the colorful displays of these indoor blooming crocus plants.

Another lovely plant that qualifies, according to some scholars, as a plant of the Bible, is the tiny *Sternbergia lutea major*. Sternbergia is native to Palestine and grows in the hills about Jerusalem and in Galilee. It blooms before the crocus in Israel today, its orange-yellow flowers contrasting with the white, pink, and pastel blooms of the crocuses that follow it in bloom, as the winter rainy season begins.

Like the autumn crocus in America, it pokes its buds up in the fall and produces delightful crocus-like flowers of a rich golden yellow. To grow sternbergia, follow the same planting method outdoors or in bowls as you would for crocus bulbs. It prefers warm, well-drained soil and does especially well in rock gardens or in crevices of old stone walls.

CYCLAMEN

Perhaps the next flower in line for the title of lily of the field, according to other authorities, is the cyclamen or so-called alpine violet. Scientifically it is *Cyclamen persicum*. This species is found even today growing freely among the rocks and in the walls throughout the Holy Land, spreading its heart-shaped leaves above the cracks and crevices. From the center springs the single, crownlike, long-stemmed blooms.

Cyclamens have an advantage over other flowers. Instead of a short blooming period, their blossoms begin in November in their native habitat and sometimes earlier, and may last a month or more. Cyclamens have been reported flowering in Upper Galilee even into early May. The ones that you can grow at home also have that happy habit of providing long blooming periods for your enjoyment.

Cyclamens in the Mediterranean Zone, to which they are well adapted, bloom through the winter with white or vivid pink flowers. The blooms are often likened to miniature butterflies poised on the slender stems of the plants. You can grow cyclamens from corms, which are available from garden centers and mail order firms. You may also obtain seed, but this requires a longer waiting period for the plant to mature. Natu-

You can enjoy modern giant mixed cyclamen plants around your home, today's version of those lovely plants of the Bible. (Photo courtesy of Burpee Seeds)

rally, you can obtain mature cyclamens from your local florist for use as a houseplant. In recent years, cyclamens have become quite popular for holiday gift-giving.

Cyclamens prefer a slightly acid soil with a pH between 6.0 and 7.0. The soil should be light in texture for young plants when you are starting from seed. For mature plants, the bed or potting mixture should be 3 parts loam, 1 part manure, and 1 part peat moss and sand. In your house, an eastern exposure is best and a temperature range between 60 and 70 degrees. During summer, keep the plants in a cool, semishaded area on porch or balcony. Potted cyclamens should be given water daily. They also welcome periodic misting of their leaves. Be certain, however, that the soil permits excess moisture to drain away.

Outdoors, cyclamens prefer semishade, regular moistening, and the soil enriched with composted humus around the plants. If you start plants outdoors from seeds, you'll find they will grow somewhat slowly but eventually will form a tiny corm. You can repot these seedlings, which sprout in 4 to 5 weeks, into 3-inch pots.

Place the forming corm so that its top is level with the soil surface. At subsequent repottings, keep the corm at the soil surface, forming a contact with the soil mixture of enriched loam and peat moss.

Few plants provide such lengthy and dazzling display of blooms as cyclamens. When blooms fade and leaves yellow, snap them off by giving the stem a sharp pull to remove the stem or leaf from the base. You can carry cyclamens over to another year by resting them in a cool spot with less moisture after they have bloomed. This provides them with a dormant period before you bring them back into their usual blooming beauty again.

During the Christmas season every year, millions of people purchase flowering plants as gifts for friends and relatives. Some of these, the poinsettia and the Christmas cactus, have no true relationship to plants of the Bible. Over the years, however, both have become symbolic of the Christmas season.

The cyclamen, however, does extend its roots back to Biblical days. During the Christmas period each year, it too is a popular gift plant. Many millions of people enjoy and admire the delicate, orchidlike flowers of cyclamens. Popularly known as Solomon's Crown, it recalls the Jewish King who built the first temple in Jerusalem. Those available from florists do still closely represent the cyclamens that can be found growing in the Holy Land today. Among the millions of people who receive them, however, most are unaware that cyclamens have deep scriptural roots.

These striking plants will bloom on for several weeks in your home if you follow several basic cultural requirements. The usual limiting factor for an extended blossom period of cyclamens is temperature. Overly high temperatures will shorten the bloom time. To enjoy the longest period of glorious blooms, place your cyclamen plant in a cool spot where the temperature will be below 60 degrees F., approximately 15 degrees C. at night.

The cyclamen we know best today, developed by the florist industry for even more delightfully full displays and larger blooms than the old, wild relatives, is intolerant of high temperatures, especially at night. Therefore, be sure to keep it away from heat registers, radiators, and other sources of heat. It is equally important, as it is with most houseplants, to keep the cyclamen protected from sudden drafts of chilly air. Cold drafts can cause premature leaf drop and permanently harm the plant.

Keeping your cyclamen away from heat also will prevent rapid drying of the soil in which it is growing. It prefers to have evenly moist but not wet soil. To further prolong the flowering period, remember to water the plant before the soil becomes too dry.

The modern cyclamen prefers bright, indirect light. You should keep it out of intense, direct light to prevent fading of the flowers. A location near eastern or western exposure windows is good, but not too close to the windows if they permit drafts across the plant.

After your cyclamen has finished flowering, and if you wish to grow it again next year, gradually reduce the watering. The foliage will turn yellow naturally and die. As many plants of the Bible, the cyclamen is a bulbous-type plant which actually grows from a corm. During the plant's growth, it manufactures food which is stored in the corm for future growth after the normal dormancy period which bulbous-type plants must have each year. After the foliage has died, remove it. Store the corm in its pot in a bright, cool place until June. A cellar window or vacant, unheated room is satisfactory, but avoid one with freezing temperatures.

Come June, repot the corm in fresh soil or potting mix that has at least 25 percent organic matter by volume. The cyclamen prefers richer soil than other type plants. Cover the lower half of the corm with the soil mixture. If you plant the corm too deeply, it tends to rot, especially if you are heavy-handed with the water supply.

Place the pot outside during June in a cool, shady location. Midafternoon shade is important to help maintain lower temperature. Maintain a moist soil condition within the pot, but be sure that excess moisture can drain away.

During the summer growing season, fertilize the cyclamen once every two weeks with a soluble fertilizer. Be certain to follow label instructions for the type of fertilizer you use to avoid over or under feeding the plant.

If you place the pot into the ground, turn it every week or two. This will break off any roots that have grown through the drainage hole in the bottom of the pot. This is necessary to prevent shock to the plant when it is moved indoors at the end of the summer. A clay pot is best for plunging in the ground. If you prefer, however, you can double-pot your cyclamen in a clay pot surrounded by peat or sphagnum moss inside an outer decorative container. This method works well for porch, balcony, or patio culture.

By mid-September, move the plant indoors to prevent any chance of frost damage. Keep your plant evenly moist and fertilize it every third week. Avoid heavy fertilizing, since that could result in leggy plants with little or delayed flowering.

By late fall, the first signs of buds on the flower stalk will appear. With proper watering and fertilizing after the summer hiatus outdoors, your cyclamen should reward you with more glorious displays again at the next Christmas season.

HYACINTH

In The Song of Solomon you'll find the most quoted scriptural mentions of lilies in the Bible. Many authorities now agree that proper botanical interpretation of the lilies mentioned there refer to the hyacinth, *Hyacinthus orientalis*. Whether you concur or elect to choose some other plant, as those before you have, you can still enjoy the fragrance and beauty of hyacinths in your gardening life.

In The Song of Solomon 6:2–4, you can read: "My beloved is gone down into his garden, to the beds of spices, to feed in the gardens, and to gather lilies. I am my beloved's, and my beloved is mine: he feedeth among the lilies. Thou art beautiful, O my love, as Tirzah, comely as Jerusalem . . ."

After reading this beautiful passage, scholars and authorities have proclaimed that it is not possible for the plant to be a lily because it is not a plant of the lowlands. Some, mistakenly looking at the common lily of the valley that grows so profusely in America, suppose that this is the lily mentioned in the scriptures. That cannot be true, because the lily-of-the-valley plant cannot be traced back anywhere near the land of the Bible.

Where authorities seem to be most confused is in trying to equate the lilies of the fields with the lily of the valley. The Goodspeed translation actually makes a point of focusing on the hyacinth in The Song of Solomon 2:1–2. The first verse reads:

"I am a saffron of the plain, a hyacinth of the valleys." Botanically (and Dr. Goodspeed undoubtedly did his scientific homework) his identification of the lilies of the valley as hyacinths seems the most logical and plausible. He goes further, in 6:3, declaring that: "I belong to my beloved, and my beloved to me, who pastures his flock among the hyacinths."

What makes many believe that the hyacinth, *Hyacinthus orientalis*, is the true plant is its wide natural range in Palestine and other northern areas of the Middle East. It flowers in the wild with deep-blue, highly fragrant spikes. Those who have visited Israel at the proper season report the hillsides in Galilee are blanketed in blue from hyacinth blooms. Through the ages, this plant has remained one of the most easily recognized natives of that area. Because of its heady perfume, the hyacinth also has been widely cultivated, across Europe and America.

From Greek and Roman days to present times, the fragrance of these lovely plants has inspired poets and planters alike. Today, you can grow deep-blue and purple hyacinths as well as pinks, reds, and whites. As bulbs, they respond well every spring when grown in fertile garden soil

The parentage of these full-bodied hyacinths, blooming bountifully and highly fragrant, can be traced to the Holy Land.

of beds and borders. Massed, these rich, perfumed flowers add an almost intoxicating fragrance to the spring air.

Hyacinths will produce lovely blooms in any type of reasonably fertile soil, provided that it is well drained. These, as other bulbous plants, do not tolerate soggy root conditions in the ground. Because these plants bloom in spring, after crocus and narcissus, they also should be planted in the fall. As perennials, they will reward you with their fragrant spikes of blooms each year. Plant bulbs between the middle of September and the end of October.

In more southern areas you may plant hyacinths into mid-November. Place the bulbs 3 inches below the soil surface in heavy soils and up to 5 inches deep in light, sandy soils. Give them sufficient room by planting 5 to 8 inches apart in beds unless you want really massed color, in which case you may plant bulbs closer together and should fertilize them during the summer after they have bloomed. Even in conventional plantings, an application of dry granular fertilizer after the blooming period will help the plants rebuild their strength for the following year's colorful displays.

You can find good hyacinth bulbs available at local garden centers. Bulbs 15 to 16 cm. provide excellent flower spikes on sturdy stems. For the largest blooms, you can order mammoth bulbs, 18 to 20 cm. in size, which are ideal for more glorious displays as well as for forcing on gravel in dishes or in water glasses.

If you prefer hyacinths outdoors, be sure to let the foliage ripen after they have bloomed. Otherwise the leaves cannot manufacture food and store it in the bulbs to nourish next year's blooming plants. This same procedure should be followed with all spring-flowering bulbs: crocus, tulip, and narcissus.

In a separate section, you'll find details for forcing many of these normally spring-flowering plants indoors during winter season. Since hyacinths offer a more unusual potential, you'll find tips on growing them alone in water here.

Top-size hyacinth bulbs perform to perfection in specially designed hyacinth-growing glasses. You can even obtain kits of these giant bulbs that nestle in the tops of these glasses. Try this blooming fun project from late October to the end of November. Before placing the bulb on the hyacinth glass, clean the root base of all old residue and roots. Fill the glass with clear, cool water until the root base touches the water. Then place the glass in a cool, dark closet or cellar until the top growth is 5 inches above the bulb. In most cases, this will occur in about 8 weeks. Keep the plants where temperatures don't exceed 50 degrees F. or drop below 40 degrees. When the hyacinth bulb has sprouted well, move the plant to a north window in your home or apartment. Add warm water if necessary, so it covers the roots that have formed in the lower portion of the container. When you place this started hyacinth in your living room, away from direct sunlight, the flower spike should grow tall and sturdy, and reward you beautifully with its giant bloom and fragrant scent for a week or more.

Hyacinths range from the blue color, similar to those found in the wild, to modern improved varieties with bright red, yellow, and a range of pastel colors.

Whether you wish to claim hyacinths as a true plant of the Bible, as a "lily" of the valley or not, the fact remains that these plants are one of the most fragrant harbingers to help you welcome spring each year.

IRIS

"I will be as the dew unto Israel: he shall grow as the lily, and cast forth his roots as Lebanon." Hosea 14:5

Although there has been much discussion and even heated argument about the true identification of the lily in various Biblical passages, this verse from Hosea seems to point more directly to another plant that is, in fact, found native to the land of the Bible.

There is such a plant in the Holy Land, the iris. Among the many types of iris botanically known to exist far back in history, and today in Palestine, the Yellow Flag, *Iris pseudacorus*, seems a most likely candidate to fit these passages. Many species of iris grow on hills and mountainsides, in fields and even drier areas. However, it is the Yellow Flag that shows its beauty at the margins of streams and waterways, often in extensive masses.

In the United States, the wild Yellow and Blue Flags also favor such environments, sometimes being swept away by floods to find another rooting place elsewhere downstream.

Throughout history, the showy beauty of irises has been greatly appreciated by kings as well as among the masses of the people. The fleur-de-lis adorns the coat of arms of ancient France and was added to one of the earlier British coats of arms. Today the fleur-de-lis, symbolizing the iris, is among one of the most often used symbols on banners of royal houses and provincial governments, such as Quebec in Canada.

Irises are handsome, showy plants and the genus contains approximately two hundred species which thrive around the world. The Eurasian Yellow Flag, *I. pseudacorus*, is one of the most common, displaying its golden-yellow blooms in June and July in temperate climates. The Blue Flag seems the best known in North America, flowering as the Yellow Flag does along streams and in marshy areas.

Because of its appealing shape, plant growers were attracted to the iris many years ago. Today, there are thousands of different varieties which provide a wide selection indeed for your gardens. From the wild yellow types, breeders have displayed remarkable ability in developing many varieties that offer every shade from light pastels to dark purples, bright yellows, and many with dazzling multicolored blooms.

Irises grow from underground stems which are called rhizomes. Since they have food storage ability in their bulbous roots, iris culture is relatively easy. Irises are classified into several groups by their root habits and flower forms. The more common types are known as the pogon iris group, more commonly called bearded iris. There are beardless types, too, including the Siberian species and Japanese iris.

Bearded Irises are readily available for your home gardens, reminiscent of the Yellow Flags whose origins go back to the Holy Land. (Photo courtesy of Parks Seeds)

You can select from these and other types, dwarf species or taller, more exotically graceful forms. Dwarf types are ideal in edges, rock gardens, and low-growing borders, since they only grow 3 to 9 inches tall. The tall, bearded types may mature 18 to 40 inches tall and are better used as background groupings or combined with other plants in formal plantings. However, groups work well as specimens in almost any garden setting. Newer types designed for American climates provide longer blooming periods, even up to early fall frosts.

Fortunately, bearded irises have a wide climate range. They can endure the heat of southern summers, as well as the intense cold of northern winters. These irises should be planted in a sunny location to perform to their peak of perfection. In shade or exposed to northern winds, they tend to be spindly and less glamorous.

You can plant irises in a wide range of soils, since they do nearly as well on heavy clay soil as sandy loams, provided the land is well drained and water doesn't linger. Even with their rhizomes planted at the surface as they should be, wet soils can be damaging to these plants. Naturally, you can incorporate generous amounts of peat moss and compost into soil to improve it for your iris beds. Although they can survive poor soils, the more you improve their growing conditions with manure and compost, the more abundantly these plants will grow and bloom. It also pays to dig beds deeply, since iris roots from their rhizomes at the surface do penetrate rather deeply in their quest for nutrients.

Plant dwarf types of bearded iris 5 to 6 inches apart if you wish a clump effect for massed blooms. Set the intermediate and taller types 15 to 18 inches apart. Place the rhizomes on the surface of well-prepared soil and cover lightly. Let the feeding roots spread out naturally and firm the rhizomes into solid contact with the soil.

Once planted, irises grow well for years. During dry periods it is best to water weekly, but never oversoak the soil. You can count on good performance for several years, but plan to lift and divide the clusters of rhizomes that eventually form. When they become crowded, lift the plants in late summer and divide the rhizomes by cutting them apart. Then simply replant in that spot, and add to your gardens in other areas with the additional rhizomes that you obtain.

After bloom, the blade-like leaves will yellow, when they have made food and stored it in their bulbs. You can cut these fans of leaves when yellowed to tidy up your beds and borders.

Japanese iris also has become popular in recent years, as have the Siberian and Oriental types. These too prefer full sun and a fairly moist, fertile soil. Divide the mature clumps of these types in early spring. Several nurseries specialize in iris plants. Wayside Gardens has a wider selection than some other mail order firms.

Irises, whether truly represented in the scriptures as some believe, or not, are certainly worthwhile additions as permanent parts of your outdoor landscape. The newer, dwarf varieties perform well in pots and in containers if they get full sun. For containers, use a rich potting mixture of equal parts humus, well-rotted manure, peat moss, and sandy loam for best growing conditions.

Irises may become infected with iris borers, and some diseases, including leaf spot and rot, may cause problems. Best protection against these problems is clean cultivation and removal and burning of dead foliage in the fall or spring. Modern combination pesticides that include insecticides and fungicides will arm you with the necessary material to control such problems should they occur.

L I L Y

"His cheeks are as a bed of spices, as sweet flowers: his lips like lilies, dropping sweet smelling myrrh." The Song of Solomon 5:13

Although much discussion through the ages has centered on the identification of the plant or plants mentioned in the scriptures as lily or lilies, this passage seems to eliminate the Madonna Lily, *Lilium candidum.* Although through the centuries, particularly following the Middle Ages through the Renaissance, attention was focused on that pure white flower, it is somewhat foolish to expect that in The Song of Solomon lips would be described as pure white.

Considering the context of this verse from the book of Solomon, we must think in terms of flowers that have a red color. Some scholars have suggested the Scarlet Lily, *Lilium chalcedonicum,* and the Martagon Lily, *Lilium martagon,* for the honor. The passage seems to warrant a plant of exceptional beauty. However, the Martagon Lily is rare in the Holy Land, and, according to botanists, it was never native there.

In the Goodspeed version of the Bible, substituting the word hyacinths in this passage also seems inappropriate. Who would extol the virtues of blue lips? The hyacinth of the land of the Bible is, of course, blue-flowered. More likely, this reference to the color of lips appropriately describes the color of anemones. These are indeed strikingly red and were found as widely throughout ancient Palestine as they are in today's Israel.

Therefore, going back to other passages from the New Testament— St. Matthew 6:28–29 and St. Luke 12:27—we must again "consider the lilies how they grow." When the verses speak of Solomon, who "in all his glory was not arrayed like one of these," perhaps new credence can be given to one species of lily as a plant of the Bible.

Whether botanists agree or not, the fact remains that through the ages, especially in recent centuries, the Madonna Lily has come to be associated with the scriptures and religious observances. Since this is the case, it is appropriate that this lovely lily be given its fair share of space in this book.

Lilium candidum, known as the Madonna Lily, is one of the most beautiful and beloved of all the lilies. It is, according to botanists, a native of southeastern Europe, blooming during June and early July.

This graceful lily was introduced to the United States many years ago and has since been grown widely. The white, glistening flowers are borne on tall, straight stalks that may tower 4 to 6 feet high. The bulbs from which this plant grows so well are base rooting and should not be planted more than 2 inches deep. Shallow planting is, in effect, one of the essentials of success.

Madonna Lilies prefer a sunny location, well-drained soil, and a slightly alkaline condition. During their growth, basal leaves are produced which remain green throughout the winter. The Cascade variety, developed in America, has large, fully open blooms which are exceptionally beautiful in form and texture.

As you attend religious services during Eastertime, you will see gorgeous lilies displayed in profusion in churches throughout the country. These seldom are the actual Madonna Lily, since more compact, new hybrids have generally replaced this old favorite. These new varieties are sturdier and, while having the lovely, pure white trumpets most people associate with this legendary lily, are much more compact in growing habit.

Madonna Lily bulbs for your garden are available from the better nurseries today. Plant them in full sun and perfectly drained soil in groups of several bulbs, each 2 inches deep. *Lilium candidum* will reward you well with some of the most beautiful white lilies the world has known.

Other varieties that also perform attractively include the Easter Lily, *L. longiflorum*, which has a similar tall form and proves excellent for garden plantings. It is frequently forced by florists for specimen pot plants at Eastertime. The Regal Lily, *Lilium regale*, also is quite similar. It prefers rich loam and full sunshine to grow most vigorously with sturdy stems covered with fine foliage. The center of this flower is flushed with yellow. It is highly prolific; a mature, well-established plant is capable of bearing fifteen to twenty blooms.

Whichever lily you prefer, each has its place in your garden to provide some of the loveliest, most spectacular displays that the lily family affords.

NARCISSUS

"The wilderness and the solitary place shall be glad for them; and the desert shall rejoice, and blossom as the rose." Isaiah 35:1

"I am the rose of Sharon, and the lily of the valleys. As the lily among thorns, so is my love among the daughters." The Song of Solomon 2:1–2

Through the years, scholarly debate has continued about the plants in these passages and at times it has not been held on the highest plane. One fact seems clear: the rose of the scriptures is most likely one of several other plants that were indigenous to the Holy Land, and not the rose as we know roses today.

Although some authorities still point to the bulbous nature of rose hips, the fruit of wild rose plants, as lending credibility to the claim that the rose of Sharon was a rose, botanical tracings refute that claim. Best educated guesses, based on tracking down old texts and both Hebrew and Greek translations, suggest the narcissus or the tulip as the "rose" mentioned in the scriptures. Most authorities now, having sifted through the earliest written scriptures and other records, agree to three possibilities. One viewpoint says the rose is a crocus. The other two schools of scholarly thought are about equally divided between tulip and narcissus. Since opinion seems so evenly divided, you'll find growing details in this chapter for both the narcissus and the tulip.

Actually, the Polyanthus Narcissus, *Narcissus tazetta*, in several similar varieties, does grow abundantly in the Holy Land today. On the plain of Sharon, a reasonably fertile area between the central part of Israel and the sea, you will find narcissi in profusion during the appropriate beginnings of the winter rainy season. These same plants also grow on the slopes of the hills around Jerusalem and of other foothills in the land of the Bible.

Along the shores of the sea, large white flowers spring up also. These are the Sea Daffodil, *Pancratium maritimum*. Closely resembling narcissi, these too are often thought to be the rose of Sharon, or even the lilies of the field. As you must realize, confusion reigns today as ever, concerning which plant is really which plant of the Bible. Siding with many others more versed in Biblical and botanical lore, we must favor the narcissus as the "rose" of the scriptures.

Narcissi, are, after all, a genus of bulbous plants which are native to the Mediterranean region. One species, *N. tazetta*, extends through Asia to Japan, attesting to its appeal to travelers, who have carried it to other lands.

Narcissus plants have long narrow leaves springing from the bulb. A

This cluster of blooms is similar to the narcissus that is native to the Holy Land.

central scape rises from this bulb, bearing one or more large white or yellow flowers which droop gracefully from the scape. The most intriguing feature of the flowers is the corona or cup which springs from the base of the flower segments, giving narcissus blooms their distinctive trumpet-like appearance.

Polyanthus or Bunch Narcissus is the type which produces many flowers on each stem. The cup of each bloom is smaller than in the more modern hybrids that have been developed for showy display. Among narcissi you have a choice of five distinct types, from the Hoop-Petticoat to the Little Group or Pheasant's-Eye narcissi. The ones that most closely resemble the natural, wild narcissus of the Holy Land are *N. tazetta* and its closest cousin, the Paper Whites.

Today, in garden centers and from mail order nurseries, you can obtain Paper White narcissi, and a similar one called Chinese Sacred Lily, another variety of *N. tazetta.* Both can be grown outdoors but do even better indoors as container-grown, forced plants. Since these varieties are closest in appearance to the wild narcissus of the Holy Land, you may wish to try them first.

You can obtain bulbs and make your own planting mix for your own container, but most firms now offer complete kits. Paper White polyanthus bulbs should be planted on top of a gravel and peat moss mix. Six bulbs will fill a dish container 6 inches in diameter. Place water in the dish as you would in forcing crocus and hyacinth bulbs, keeping it resupplied to the level of the base of the narcissus bulbs. Within weeks, you'll be rewarded by large, sweet-scented clusters of white flowers with their diminutive light-yellow central cups.

The Chinese Sacred Lily is similar to the Paper White Narcissus, both in growing habit and in culture. However, the center of this narcissus is bright yellow, which contrasts nicely with its cream-white petals. You may also select other types of narcissus for dish- or pot-forcing indoors. The golden-yellow variety, Soleil d'or, is also similar to Paper Whites, but its flowers are deep yellow, with an elegant orange-yellow cup.

Other types of narcissus and daffodil, their kissing cousins, can be made to bloom indoors as well. Plant the bulbs of these large specimens in potting soil or peat and vermiculite mixtures as the florists do. Keep the soil mixture moist and the container in a room of 48 to 50 degrees F. for 5 to 6 weeks. Then, when the sprouts of leaves and flower stalks are 4 to 5 inches tall, bring them into a cool room, 55 to 60 degrees F., to watch them burst into glorious bloom. If you wish to force narcissus, it is important to buy the kits or the specially selected variety of large bulbs. These are offered by nurseries, reputable mail order firms, and reliable garden centers to insure your success with indoor-blooming bulbs.

This handy bulb planter lets you make a hole exactly the right size for planting bulbous hyacinths, narcissi, and tulips in your outdoor garden.

Outdoors, plant your narcissus bulbs at least 4 to 6 inches deep in the fall, between September and mid-November, or earlier in northern states. Poke a hole in the ground with a dibble or open the earth with a trowel. You can use special cylindrical planting tools for easier setting if you plan to deck your landscape with dozens of narcissi. Be sure the base of the bulb is firmly in contact with the soil. Although some people enjoy these plants in beds and borders, they seem most attractive when planted randomly in naturalized woodland settings.

STAR-OF-BETHLEHEM

"And there was a great famine in Samaria: and, behold, they besieged it, until an ass's head was sold for fourscore pieces of silver, and the fourth part of a cab of dove's dung for five pieces of silver." II Kings 6:25

Growing in the wilds throughout the Holy Land, the star-of-Bethlehem dots fields, hillsides, and other stony places with its bright, white blossoms each year. It seems ludicrous to consider that this plant was the "dove's dung" described in the book of Kings. Some scholars believe that the phrase has no relation to any plant, much less this lovely, delicate one. However, as other researchers have studied the scriptures and traveled the Holy Land, they have been amazed at the multitudes of these tiny white-starred flowers which burst forth in such profusion with the first rains of the winter rainy season.

As these authorities have seen and studied the whitening of the rocky hillsides, it seemed to them that the blooms of the star-of-Bethlehem dotting the hillsides were much like the whitening of the ground and buildings from the dung dropped by doves and other birds. In this symbolic way, the plant may actually be a true plant of the scripture.

In early botanical texts, as early as Linnaeus, this plant was reasonably accepted as the dove's dung of scriptures and in fact today the common name for *Ornithogalum umbellatum* is indeed dove's dung. In other writings dating back nearly to the time of the Bible, it has been stated that bulbs of this plant were commonly collected, ground into meal when dry, and mixed with flour to make a type of bread. Such records are, however, questionable, since modern scientific tests show the bulbs to be toxic. That fact may seem to disqualify this plant as one of the plants of Bible times, but all other factors, including its presence there today, lean toward the positive view.

Since most botanists now seem willing to accept the interpretation that the star-of-Bethlehem is most likely the "dove's dung" of the verse in II Kings, we have included this lovely flower here so you may grow it in your garden with the others from the scriptures.

The star-of-Bethlehem, *Ornithogalum umbellatum*, is a bulbous plant that grows 6 to 12 inches tall. It has many narrow, grasslike leaves which have a white band along their midrib. The numerous white flowers are all star-shaped, carried on loose umbrella-like stems at the top of a naked stalk.

You can grow these plants easily from bulbs. Plant them in reasonably fertile soil, 2 inches deep. They do best in natural settings, beneath trees and shrubs that put out their leaves after the star-of-Bethlehem has bloomed. Plant your bulbs in the fall, the same time you set tulips, nar-cissi, and crocuses. Come spring, the daily, free-blooming star-of-Bethle-hem will delight you with its showy clusters of white flowers lined under-neath with white stripes on each leaf.

T U L I P

"I am the rose of Sharon and the lily of the valleys." The Song of Solomon 2:1

Earlier in this section you read about the debates that have been waged concerning the true identity of the rose in the scriptures. Although some still favor the autumn crocus, and some adhere to their belief that it is the narcissus, there is a group of scholars that is adamant that the rose in these scriptures really is the tulip, *Tulipa montana.* Among these schol-ars, some lean toward the *Tulipa sharonensis.*

To refute the belief that the rose could truly have been a rose, some botanists have pointed out that there are no true roses in the climate of the Holy Land. That is not exactly accurate, since some old-fashioned wild brier roses are known in the mountains of Lebanon.

Again, the translations of Drs. Moffatt and Goodspeed add confusion to the mystery. Dr. Moffatt translates The Song of Solomon, "I am only a blossom of the plain," while Dr. Goodspeed's version translates, "I am a saffron of the plain, a hyacinth of the valleys." Both these learned men have their botanical and Biblical adherents.

More recent research, tracing the plants of the land of the Bible, has been done at the Hebrew University in Jerusalem and elsewhere, which again focuses on the attractive bulbous tulip plant with its scarlet flowers as the rose. Since this species of tulip is commonly found in the mountain-ous regions of Lebanon and Syria too, it may indeed be the rose of the Song.

From ancient records of other civilizations, we know that tulips were cultivated ages ago in Turkey. The name tulip, some linguists believe, derives from a Persian word meaning turban. Whatever you believe, after reading through the scriptures and various translations of the holy book, there is no doubt that tulips do grow wild even now in modern Israel. Since they do and they are such lovely flowers, it seemed appropriate to include them in this book.

Tulips have become a multimillion-dollar horticultural industry. Holland has long been famed for its production of these flowers, exporting

These sparkling tulips along a natural old stone wall remind us of the vivid tulips that grow wild even today in the land of the Bible.

hundreds of millions of bulbs annually. In the 1600s a wild tulip craze began in Europe, with collectors paying hundreds of dollars for single bulbs of newly developed, more decorative strains and varieties. Today, you have a wide selection of colors and styles; in fact, thousands of different varieties and types. No doubt, most are descended in one way or another from the tulip that was perhaps the rose of the scriptures.

For your garden, here are the best cultural methods to enjoy the best of what the tulip world offers.

The wild tulip of the Holy Land bears a single, bright red flower. Today, you can quite closely duplicate the beauty of this plant by carefully selecting such varieties as the blazing Red Emperor, Red Riding Hood, or Scheherazade. True, these are modern tulips, but they are representative in shape and form.

Greigii species and hybrids offer you large flowers as well as attractive foliage. Related to the wild Oriental *Greigii* tulips, these lovely specimens do well in good sun to semi-sun. Species tulips are particularly valuable because they multiply and require little care. They bloom very early and are especially attractive in rock gardens, along walls, or in naturalized settings. *Praestans* seems remarkably close to the appearance of the wild tulip of the Holy Land. *Princeps* also has a brilliant scarlet color which blends well with other plants in low beds and borders.

You may, of course, prefer the Lily-flowered, the Kaufmanniana or Water-lily tulips, Fosterana hybrids, or other fancier varieties. That choice is yours, remembering that the tulip thought to be a plant of the Bible is notably more wild and less showy than these intensely bred new varieties.

Plant your tulip bulbs between early October and mid-December. If you plant too early, they may begin active growth and can suffer a setback during winter. It is better to wait until after you have planted your narcissi before setting tulips. They must be in the ground well before it freezes, however, if they are to set firm roots for spring growth.

Ideal soil for tulips is a light, fertile, well-drained loam. Never use fresh manure on tulip beds, but well-rotted compost is a good addition to improve growing conditions. Tulips produce their best blooms the spring following planting. Some varieties actually bloom well for several years, but others deteriorate rapidly and must be replaced. The species tulips are hardier, since they require little care and multiply themselves to add to your gardens' loveliness year by year.

As with narcissi and hyacinths, leave tulip foliage alone until it finally yellows. Foliage is the food-building part of these plants and must make and store the food within the bulb for next year's performance.

In cold northern areas, it pays to mulch over tulip beds in the fall, covering them with straw or old dried leaves.

Gardeners who live in southern areas often have fretted that they could not enjoy the glow of tulips around their homes. It is possible. By refrigerating tulip bulbs for several months, then planting in December to early February, you will have forced a dormancy on the bulbs. They'll begin growth as soon as planted, and you may be amazed how well these plants will perform, even in southern areas of the United States.

Mail order catalogs list such a profusion of new varieties and old favorites; it is almost impossible to name those which are best for your needs. If you request the free catalogs listed in this book, and those you see in horticultural and gardening publications, you can take your pick of the best that the garden industry has to offer.

CHAPTER VII

VEGETABLES
OF THE BIBLE

"We remember the fish, which we did eat in Egypt freely; the cucumbers, and the melons, and the leeks, and the onions, and the garlick." Numbers 11:5

You too can enjoy tastier living when you grow fresh vegetables in your home garden. As the children of Israel hungered for the flavorful vegetables they so fondly recalled from their time in Egypt, the thoughts of the goodness of those vegetables was strong in their memories. Today, nostalgia, a longing for the better times of the "good old days," is a recurrent theme across America.

During the past few years, many millions more people have begun to plant and tend backyard vegetable garden plots. Family gardening has become America's number one hobby, but not necessarily for the reasons you might at first suspect. It is true that millions of families, faced with climbing food costs, decided to begin gardening, or to expand their previous plots to help fight inflation and reduce their food bills. Others, of course, garden for the fun of it.

When you ask people today why they continue to plant and tend their vegetable gardens, their answer is more likely to be "for the flavor and the fun of it." In this respect, modern Americans are saying to themselves and their neighbors what the children of Israel had said so long ago in the land of the Bible. Flavorful food has a way of impressing itself on our minds.

As those in the land of the Bible remembered so vividly, and discovered again, as they began to till the earth and sow their crops, the taste of homegrown vegetables is most flavorful. There are several reasons for this, most veteran gardeners know.

First, the vegetables are freshly picked. Although today, with rapid

You can grow cucumbers right in a pot on your patio or apartment house balcony with these dwarf-size plants that yield abundantly. (Photo courtesy of George J. Ball)

transportation from farms to supermarket shelves, much of the freshness of produce can be preserved rather well, especially with modern refrigeration. However, no truck or even airplane can match the speed with which you pick your vegetables from your backyard garden plants and rush them to your kitchen stove.

There is another reason, which explains the tastier nature of home-grown foods, that may escape some people. When you grow your own, you can carefully select the vegetables at the peak of perfection, when they are perfectly ready and ripe. If some are overgrown or overly ripe, you may discard them.

Perhaps the most important hidden reason that accounts for the more delicious taste of homegrown vegetables is a factor that is not understood so well by most people. Commercial farms must have varieties of garden vegetables that respond well to intensive cultivation, that yield heavily, and mature with uniformity for mechanical harvesting. That must also hold up well during harvest, handling, and shipping to stores across the country.

Many superior-tasting vegetable varieties do not meet these require-ments of commercial farmers. However, you can easily grow the more flavorful varieties that do perform abundantly well in home gardens. You will find that they will not only be tastier but also yield over longer periods, and provide you with rewards far beyond what you can find in any supermarket.

Among the plants of the Bible, cucumbers and melons in particular, you have a wide selection of modern varieties which closely resemble those that had their roots in the Holy Land. Equally important, thanks to the creative talents of botanists and plant breeders through the years, you can grow these vine vegetables even in small plots or even tubs and pots. New cucumbers and melon varieties have been perfected that grow as bushes, rather than long, trailing vines. As God helps those who help themselves, so are we helped by the dedicated work of plant breeders who have used the genetic pool of old-time varieties to develop improved plants for our use today.

In this chapter, you'll discover some old-time favorites and their newer relatives, all from the same families as those plants grown in ancient Palestine. By applying your talents to the soil, you too can savor the dis-tinctive flavor of these vegetables in abundance.

CUCUMBERS

"And the daughter of Zion is left as a cottage in a vineyard, as a lodge in a garden of cucumbers, as a besieged city." Isaiah 1:8

Cucumbers are annual trailing vines and are among the top ten most popular vegetables in the United States. Many of the varieties available today also will climb a fence or trellis. More recently, dwarf bush varieties have been introduced which enable you to grow these age-old vegetables in a tub or large pot on porch, patio, or even a balcony high above the city streets.

These tasty vegetables have been cultivated in all the warmer climates of the old world so long that their origin has been lost. It is known, however, that cucumbers were an important part of the diet of people, especially the poorer people, throughout Egypt and other countries of the Mediterranean area for centuries before the life of Christ. The "lodge in a garden of cucumbers" refers, according to Biblical scholars, to small houses often built in fields and vineyards in ancient Palestine for shelter during the heat of the day.

Tracing the roots of these fruitful plants into antiquity, it is possible to say with certainty that they were extensively cultivated in Egypt along with other common foods of the masses—onions, leeks, and melons—thousands of years ago.

In fact, cucumbers still are widely grown there and in Palestine today, and, together with breads, are eaten raw, as well as pickled for storage.

You have a wide choice of cucumber varieties from which to pick the type that will please you best. Some have been developed for eating fresh, cut into spears, or sliced for salads. Other varieties, also descended from the cucumbers of the Holy Land, are better for pickling into the traditional dill pickles so favored by many people. Dill, of course, is one of the herbs that also is mentioned in the scriptures.

Happily, newer, improved varieties have been bred which solve some of the problems gardeners have encountered with older-type cucumbers. New disease-resistant varieties defy disease such as the wilts, which can kill nonresistant vines. Others, like new Bush Crop, produce prolifically on a compact bush plant with very short vines. Even if you live in an apartment, you can grow these in pots. The attractive, full-size fruit matures 6 to 8 inches long and has fine-tasting, crisp flesh.

Cucumbers prefer a sunny location. Considering their native habitat in the hot weather of the Holy Land, that seems natural. They also prefer rich, well-fertilized soil to perform best. If your soil is poor, mix in compost, peat moss, and manure. Improving the soil also helps it to retain moisture, which is especially important for cucumbers. One secret of successful cucumber-growing is ample water. They, like melons, become terribly thirsty. Be sure the soil never becomes dry, especially during the fruiting time.

Plant cucumbers in late spring when the soil is thoroughly warm. You can start seeds in peat pots indoors several weeks before setting them outdoors. This will give you a jump on spring. However, cucumber seedlings are sensitive to frost, so never plant them outdoors until the danger of the last spring frost has passed. Remember, too, that even the large-size

Cucumbers were a favorite food in the days of the Bible. Today, you can grow tasty, crisp cucumbers in your home vegetable garden. (Photo courtesy of All-America Selections, Los Altos, California)

cucumber seeds can rot in the ground if the soil is too cold and wet. If you live in northern areas with short growing seasons, take heart. Newer, quick-maturing varieties are available that let you enjoy the delights of cucumber eating, even in northern areas.

Since cucumbers are exceedingly prolific, you will not need many plants to supply your needs all season. Even better, cucumbers have a natural tendency to produce more fruits the more you pick those that mature. Keep picking and you encourage greater productivity from each vine or bush.

Hill culture is the best growing method. For the vine-type trailing cucumber varieties, either plant in rows 6 to 8 feet apart and space seeds 4 inches apart, or space hills 4 to 6 feet apart and place 4 to 6 seeds a few inches apart in each grouping. Cover seeds one half inch deep and firm the soil over them.

When seedlings are about 2 inches high, thin them to 3 plants per hill, or, if in rows, to 12 inches apart. As you thin to prevent crowding, also plan to mulch the hills or rows. Apply several inches of compost or old leaves around the plants. This helps smother weeds, absorbs the sun's warmth, and also helps retain soil moisture, which cucumbers need.

Depending on the variety, cucumbers will be ready for your table within 60 to 75 days. This will vary by degree days in your area. If you want tender young cukes, pick when they are small. Or, you may wait a day or so and pick them at full size. When cucumbers begin forming, check your plants daily. These fruits grow rapidly, practically before your eyes. If you aren't observant, they'll pass their prime and become somewhat bitter, with too many large seeds.

Among the recommended varieties are Sunnybrook, which has few seeds, is hardy and a good all-purpose cucumber. Spartan Valor hybrid is an All America selection with vigorous vines that produce many fruits. It is highly resistant to mosaic and scab disease, too. M & M hybrid also is disease-resistant, sets fruits well, and has a fine, mild taste. Triumph hybrid cucumber and Bush Crop are excellent if you have little garden room. Some people find cucumbers a bit hard to digest. If they disagree with you, perhaps you should consider Burpless. It is an aptly named, easily digestible variety worth trying.

MELONS

" . . . which we did eat in Egypt freely; the cucumbers, and the melons . . ." Numbers 11:5

During the searing heat, while the tribes of Israel wandered in the desert, thoughts of the cooling refreshment of the melons that they had enjoyed in Egypt filled the minds of many, as the scriptures tell us. Who today can resist the tantalizing aroma of ripe melons, especially during the hot days of summer?

There are those who have studied the scriptures who say categorically that the melons mentioned in the passage from the book of Numbers are truly the muskmelons, *Cucumis melo*. Almost as many authorities believe that they are instead the watermelon, *Citrullus lanatus*. Either one or both could be, in fact, the melons which the children of Israel thought about so longingly in the desert of the Sinai.

Although opinion is divided still, it is more likely that the word "melons," translated from the early Hebrew word "avatiach" or "avatichim," refers to both types of melons. From time immemorial, both melons have been known in ancient Egypt and the areas of the earliest civilizations of what is today the Middle East. Among botanists, there is a belief that the muskmelon, also called cantaloupe today, is a native of Egypt and the Levant. Others place its origin variously in India and central Asia.

As caravans traveled from afar, long before the time of Jesus, they carried more than mere goods of trade. No doubt, as research into ancient history has proved before, caravans carried seeds of plants and often rootstocks as well. No matter that the true roots of the muskmelons may have sprung in other lands, the fact remains that these fruits had been cultivated and enjoyed by the pharaohs and the people of Egypt since long before the first written records were set down.

Scientifically, the muskmelon is a tender annual trailing herb, with palmately lobed leaves. It bears tendrils by means of which it readily climbs over fence and trellis. Many people consider melon a fruit, rather than a vegetable, but scientifically it is classified as an herb. As you read the scriptures, you find no reference to vegetables but many passages about herbs.

Although today we may think of herbs as those plants used for flavoring and spices in our other foods, many plants are actually herbs. It would seem, therefore, that among the references to herbs in the Bible a broader meaning may have been implied, encompassing foods like melons as well as others that we think of today as vegetables or fruits.

Muskmelons, native to ancient Persia and southern Asia, won fame among many peoples of the ancient world. The Romans and the Greeks were familiar with them. Old records show that melons were extensively cultivated in France by the early 1600s. Even today, muskmelons are popular in northern Europe, where they must be grown under glass in greenhouses that provide the heat they need to produce abundantly.

The watermelon also has been cultivated in the land of Egypt since the earliest written records of history, and probably long before, too. Botanists believe this fruitful vine is a native of tropical central Africa, but whatever its true origins, it long has been a valued food in Egypt as it is today. Watermelons prefer warm weather, and if you were to visit Egypt now, you would find that these large and luscious fruits are an important part of the diet of the poorer people. They also are a delight for the richer citizens there.

Travelers to Egypt have remarked how curious it seemed that watermelons are depicted in ancient carvings and drawings. That, no doubt, attests to their popularity far back in time during the lives of the early pharaohs.

Watermelon plants are prostrate, long-running, hairy vines with branched tendrils and large leaves, somewhat oblong in shape and deeply cut into several narrow lobes. Typical fruits are light to dark green, often striped with light and dark lines.

It may be interesting to note that the watermelon varieties which we regard as best today, the sweetest with the fewest seeds, may not find favor elsewhere. In Egypt in this modern age, watermelon seeds still are valued. They are saved to be roasted, salted, and eaten as a popular dish, much as many people in the United States find sunflower seeds a prized treat.

As farmland has been restored to more productive capacity in Israel today, melons again are a major crop. The revived agricultural areas between Jaffa and Haifa can produce melons of incredible size and delicious taste.

You can enjoy both muskmelons and watermelons in your home gardens. No longer are they restricted to parts of our country that have long, hot growing seasons. Now, new hybrids, perfected from the best cross-breeding efforts, enable you to grow both types of melons in the shorter seasons of somewhat cooler northern climates. More good news. Newer varieties that have been introduced do not trail long vines. With limited space or even a large tub on your apartment balcony, you can appreciate the fruits of what some writers call the ambrosia melons among melons.

Muskmelons prefer warm, rich soil. Generous additions of organic matter help improve their growing conditions. You should also apply ample fertilizer to speed their growth so you achieve the delicious, desired results, the tastiest melons on your vines.

Because these melons and others can't abide cold weather, wait until your outdoor soil is thoroughly warmed in spring. Best culture for cantaloupes is the hill method, as it is for garden cucumbers or squash. Prepare several hills by mounding the soil slightly, leaving a central depression. Place 4 to 6 seeds in this 6-inch diameter, depress and cover with soil, firming it well.

Water weekly, if rains don't come, so seeds sprout properly and get rooted well. If you choose the trailing vine varieties, mulching with leaves, grass clippings, old straw, or compost serves to preserve soil moisture, as it smothers weeds and saves you hoeing work.

All types of melons are composed primarily of water. That means

Smaller, faster-growing watermelons like this Honey Island, developed in Norway, thrive in northern gardens.

they require vast amounts of water in the soil as the vines grow, flowers form, fruits set and mature into prized melons.

In growing these tasty melons, follow the cultural directions for cucumbers. They enjoy the same conditions: warm days, lots of refreshing water regularly, at least each week, and protection from the competition of weeds that rob moisture and nutrients. You may find with melons and with cucumbers that black plastic is a valuable aid. These sheets of black polyethylene film can be laid on the ground, their edges and ends held in place by soil. Simply plant seeds through slits in the plastic. As seeds sprout, the plants rise above the plastic, which smothers weeds and absorbs the sun's heat to speed the plants on their path to maturity.

Varieties may be trailing, semibush, or dwarf growth, so follow the spacing recommendations on the seed package.

If you are digging into the field of container gardening, you still can enjoy the tasty treat of muskmelons. New varieties yield personal-size fruits, on bushy plants which thrive in tubs and buckets. In these, of course, you must pay more careful attention to providing water, at least three times a week, since containers on balconies or rooftops tend to lose their moisture in hot summer sun.

Liquid fertilizer applications, twice a month during the season, also will be necessary to insure that your melon plants get their fair share of nourishment. Remember that you must feed the soil to feed the plants, or most assuredly they will not respond in kind to feed you well.

Some older varieties may yield well but unfortunately may be prone to plant diseases that infect melon vines. If diseases are a problem in your area, select the new disease-resistant varieties.

One perennial question about growing melons is: When are the melons ripe enough to pick? Some gardeners tap while others tug to test the melon's ripeness. Some new varieties can be identified when they are ready by the coloration. However, the most reliable way to tell when your melons are ripe and ready for tasty eating is to give the melon a light tug. If it separates easily from the vine, it is most likely ready to be picked, since it comes away in your hand. If it resists your pull, give it another few days to ripen fully.

Thanks to plant breeders who have tapped the genetic pool of muskmelons dating to before recorded history, you have a wide selection of fine muskmelon varieties to plant.

Sweet'n Early hybrid is an early maturing treat with firm, bright, salmon-colored flesh that is juicy and sweet. Vines are resistant to powdery mildew and bear 6 to 8 fruits per plant.

Iroquois, developed by Cornell University, is one of the tastier home garden varieties. Fruit is moderately netted, the seed cavity small, and this one is resistant to troublesome fusarium wilt. Hearts of Gold has thick, deep-orange flesh of deliciously sweet taste. The melons are 6 inches long.

Honey Rock and Samson hybrid combine disease resistance to wilt and powdery mildew with heavy sets of fruit. Both are delicious, too. For large size, Burpee Hybrid wins applause. The new Ambrosia hybrid with luscious, light salmon flesh that is firm, very sweet and juicy is a relatively new introduction. Vines are disease-resistant, seed cavity tiny, and the flesh richly thick and tasty.

If you live in northern areas, try Mainerock hybrid, ideal for northern gardens, bearing excellent fruit early. Minnesota Midget uses less space than other varieties, with vines growing only 3 feet long, yet this tasty muskmelon yields delicious 4-inch melons that are perfect for individual servings.

WATERMELONS

Many gardeners have bypassed growing watermelons because they believe their season is too short and that they lack the room to accommodate sprawling watermelon vines. Today, you can cultivate these refreshing melons, these plants of the ancient Holy Land and the Bible, even in small, northern gardens.

It is true that most of the commercial watermelons that arrive at supermarkets are grown in more southern climates, primarily in Texas, Arizona, and California, but also across much of the southern farming belt. Most likely you have enjoyed watermelons often, as did people who lived thousands of years before you. Until you try some of the superior new varieties that thrive in home garden grounds, you haven't eaten the best that watermelons have to offer.

If you cultivate other crops in your garden, such as corn or pole beans, you can intercrop with watermelons. Their vines will ramble along the ground from the hills in which you plant the seeds. Since the plants are nourished through their roots in the hills, double-cropping makes good sense, letting you double the productivity of your land. Remember, however, that whenever you plant extra crops closer together, the land will need extra fertilizer and water to support this multiple cropping.

Watermelons, like muskmelons, prefer lots of sun and warmth. Plant seeds after all danger of frost is gone and the soil is warm. Well-drained loam is best. If you have poorer soil, you can improve the immediate growing conditions for melons by enriching the soil of the hill in which you plant the seeds. Dig down a foot in a 24-inch-diameter hill. Mix the soil with compost, manure, and peat moss if it is sandy or a heavy clay.

Sow 2 to 8 seeds per group or hill, and thin to the strongest three plants. It pays, as in all garden crops, to select only disease-resistant varieties. That bonus has been built in by plant breeders to insure greater satisfaction to those of us who garden. Some old-time watermelons are susceptible to wilt and anthracnose disease.

For your watermelons, follow the cultural procedures you should for muskmelons: water regularly—lots of it—and keep the hill areas clear of weeds. Mulching is a valued practice.

The best varieties for most gardens are those that require less space. If you wish to emulate the watermelons of the Holy Land and Egypt, however, try Charleston Gray. This variety combines fine eating quality, resistance to disease and sunburn. Its flesh is crisp, red, and delicious and melons mature between 25 and 35 pounds. It does need room for its vines to roam. Since most of us have limited space, Burpee's Sugar Bush has advantages. It requires only 6 square feet of space. Plants have short-branched vines only 3 to 4 feet long, but they produce several melons up to 8 pounds each, a convenient refrigerator size. Sugar Baby, an early maturing type, has sweet, fine-textured red flesh with fruits 8 inches across. Sweet Favorite hybrid matures in only 80 days and is a valued All America award winner. These vines are disease-resistant and produce large melons up to 20 pounds with bright scarlet, sweetly juicy flesh and small seeds.

New Hampshire Midget is another All America award winner, which means it has proved productive in all parts of the country. Since it is very early and matures in only 70 days, it is designed for northern gardens. Fruits weigh up to 6 pounds each and the strawberry-red flesh is solid and sweet. Although this variety requires little space, its vines are vigorous and heavy-yielding.

Each year, plant breeders continue their work, crossing the best available varieties with others to achieve still greater yields, earlier harvests and sweeter tasting fruits. These are just a few of the better musk and watermelons which you can grow this year. As you do, and as you eat them, the longing of the children of Israel for their melons will be more understandable. Melons have been a favorite food of mankind since they first took root.

ONIONS

" . . . and the leeks, and the onions, and the garlick." Numbers 11:5

In the time of privation and hunger in the desert, the children of Israel may often have wondered whether the abundance that grew in Egypt would ever grow in the lands they wandered. During this time of dissatisfaction and near rebellion against their leaders, many of the people most likely dreamed fondly and often of the foods they had enjoyed in Egypt, despite their bondage there.

One of the foods mentioned in the scriptural passages of this era is the onion. It, like the cucumber, was and is a staple food in the Middle East as well as in Egypt today. In fact, the fame of the Egyptian onion which grows in the rich soil of the Nile River delta extends far beyond the borders of that nation.

You may think of onions as a harsh or bitter vegetable. The onions so well known and widely grown in Egypt and other parts of the eastern Mediterranean are far different from the hard-coated types. They are, as are the popular Bermuda onions many know and grow today, sweet and mild. In Biblical days, onions were a basic food since they could be eaten fresh, cooked, or dried for storage.

In the Great Pyramid of Cheops there is an inscription which tells of the price paid—many talents of silver—to buy the onions and other food for the workers while the pyramid was being built. One scholar estimated that, at today's valuation, more than three million dollars were spent for onions, garlic, and other root-type crops.

Egyptians were so fond of the onions that one was actually accorded divine honors, according to inscriptions in ancient tombs. The edible part of the onion is the thickened base of its leaves that rise from an extremely shortened stem. The small, hard structure at the base of the bulb is the stem plate. Upper portions of the leaves are cylindrical and hollow. The seed stalk is formed during the second season and, depending on the type of onion, may be 2 to 4 feet tall, topped by a globe-shaped flower. You may have seen this flower in some gardens, since *Allium* is available in a flowering variety.

Onions, even the Egyptian and Bermuda types, have a characteristic pungent taste and odor. Botanists, delving into the lore of Biblical plants, believe that the onion of the scriptures is mostly like the *Allium cepa*. Through the ages the onion has retained its unique characteristics. Although today it is still used as a food in its native lands, it is more commonly favored elsewhere as flavoring in soups and stews, in salads, and as accents to other foods, rather than a food itself.

Actually, the onion is a vastly underrated vegetable. Forget the often bitter onions that you find in supermarkets as you plan your garden. You can grow onions that are sweeter, juicier, and far more appealing if you wish. Every garden should have a place for them.

Onions prosper in almost any soil, but as you improve their growing conditions, you'll be rewarded with larger onions, since they respond so well to extra care. Adding organic matter to the soil improves their growth. This is understandable, considering how they thrive in the rich soils of the Nile delta and along its banks, which are replenished periodically at flooding time.

Today, you have a reasonable choice of varieties, as well as planting methods. You can plant seeds, plants, or onion sets. Sets are the tiny bulblets grown by seed firms from seeds, then dried and sold for easier, faster planting and more satisfactory, uniform results in most garden soils. Your choice of variety is wider if you grow from seed.

An onion's natural tendency is to grow its tops in cool weather and produce its bulbs in warm weather. So, plant onions as early in spring as your soil can be worked. It should be well warmed and of a proper, crumbly texture. Place the sets, which you can usually buy locally or from mail order firms, in one-inch-deep furrows. You can plant double or triple rows, providing you space the sets to give each plant room to grow without crowding. Merely press the sets into contact with the soil in the furrows and cover them lightly with soil. Sets may rot in wet weather, and really only need firm soil contact to sprout and root. Space sets 3 inches apart and rows 12 to 20 inches apart. If you use double rows, place sets 3 inches apart, zigzag fashion. A pound of sets will produce 50 feet of single row, 25 feet of the zigzag fashion.

In recent years, another good idea has been transplanted to America from Europe. It is called the French intensive gardening method. This system employs wide rows in which many seeds or plants are placed more closely together. With this cultural technique, you get much greater yield from each square yard, but you must provide additional nutrients and water to produce the abundant yields you can expect.

Onions also can be grown rather easily from seeds. Relatively few people try this method, since sets are so readily available and easier to plant. However, your choice of tastier varieties is much better from seeds. You also gain a bonus: young green onions for salads as you thin the rows of onion seedlings. A packet of seeds will sow about 20 feet of row.

Using seed, sow early, covering the seeds ever so lightly with a quarter inch of fine soil or sand. Whether you grow from seeds or sets or prestarted plants, clean cultivation, removing weeds regularly, is important. It only takes a few minutes each week, but this little extra effort greatly increases the speed of maturity and quality of your crop.

When bulbs have reached maturity, break off the tops, the hollow leaves, to hasten ripening. It is best to break off any flowering stalks that form too. After a week or so, as the onions ripen, you can pull them up. Leave them on the ground in warm, dry weather, or bunch them together by tying the tops and hang them against a wall to dry. You may also hang them over the beams in a garage, out of the weather so rain won't soak the ripening, drying bulbs. Most onions store well when hung to dry.

Since onions do require a fairly long growing season of warm weather, you should consider seeding in flats indoors before outdoor planting time after the last spring frost. Some firms also offer onion plants by mail which are usually hybrid varieties which mature in about 65 to 70 days from planting. From seeds, most types take 100 to 120 days to harvest.

Try these onion varieties for the best eating, as you capture some of the flavor of the ancients from Egypt and the land of the Bible.

Sweet Spanish onions are close kin to the large onions eaten so widely by the people of the Holy Land. (Photo courtesy of Burpee Seeds)

Sweet Spanish is a very large, globe-shaped, fine-grained, and really sweet onion. It has yellow skin with pure white, firm, and crisp flesh. The mature onions range from 4 to 6 inches across and are excellent raw or cooked. A white Sweet Spanish also is available with white skin, but this and the yellow type do not store well.

Yellow Bermuda is extremely mild with juicy, white flesh. These are medium-sized, somewhat flattened onions with a straw-yellow skin, good for southern areas but not recommended for above the Virginia climate zones.

Yellow Globe hybrid is an outstanding one for early maturity. These onions are mild-tasting, 3 to 4 inches across and keep well. Crystal White Wax is another early, Bermuda-type onion which is sweet and surprisingly mild-flavored. Onion bulbs mature medium to large with white skin and waxy white flesh. These too do well in southern areas but are not recommended for northern gardens.

A more colorful onion and perhaps more appealing in appearance, Giant Red Hamburger is a hybrid that is tasty for slicing for sandwiches and salads. The large, semiflat onions have red skin with red and white flesh.

Once you tune up your tastebuds to onions, you may prefer more oniony onions. Try Yellow Globe Danvers. This is a productive variety that keeps well and has a strong, hearty flavor from its light golden, firm, and uniform bulbs.

To those who have cultivated a taste for and a growing ability with onions, these vegetables add zesty flavor to your meals. You can use them in salads, soups, sauces, and sandwiches or prepare them as a hearty bowl of onion soup.

As a basic food that fed so many people for centuries before the life of Christ, onions have a rich heritage among the plants of the Holy Land.

LEEKS

The climate in the Holy Land, consisting as it does of two seasons, through the ages favored the development of bulbous-type plants. Many of the flowers that developed there and managed to survive the trials of the centuries, grew from bulbs. The same, to a degree, can be said for some of the vegetables. Onions are one example, leeks another.

Debate still continues among botanists with some believing the leeks mentioned in the scriptures are truly leeks as we know them, *Allium porrum*. Others state that the leeks of the Bible are more likely a leguminous plant known as fenugreek, with a Latin name, *Trigonella foenum-graecum*. The fact remains, both plants were known ages ago in Egypt and the land of the Bible and were popular as food there aeons before the birth of Christ.

Closer study, however, leans to the decision that the leek of scriptures is the leek we know as the true leek today. Related to the onion, the leek has a bulb which is more slender and cylindrical. Its leaves are flat, somewhat broad, and more succulent than onion leaves, which are seldom eaten by most people. Although the flavor of leeks is similar to that of an onion, it is judged more pungent.

In Europe and the Middle East today, leeks are traditionally used for seasoning for stew, soup, and meat cooking. The leaves, however, are widely enjoyed in salads and chopped for flavoring of soups and stews.

You can grow bunches of leeks which also are mentioned in the scriptures. (Photo courtesy of Burpee Seeds)

Leeks are probably of eastern origin, since they were commonly cultivated in Egypt in the time of the pharaohs, and are to the present day. They were known in England by the latter part of the sixteenth century, probably brought there by the Romans. The Roman emperor Nero was said to have held leeks in great esteem, eating them every month. Behind his back he was called by his detractors "eater of leeks."

If you enjoy gourmet cooking, leeks belong in your garden. In England, leeks are used primarily in soups and stews, except for the bulbous roots. In French cookery, the blanched stems are employed widely.

Leeks grow much like onions, with sheathing stalks of the leaves lapping over each other to form a thick, stemlike base. When soil is mounded around this part of the plant, the stalks blanch naturally.

Leeks are relatively hardy and present no difficulty in cultivation. However, in the colder parts of the United States, these plants should be dug and heeled-in during winter. Some gardeners prefer to winter them over in pots in a cool, dark basement. These plants are biennial, flowering the second year of their growth.

As you read the Bible references to bitter herbs and grass, you may feel, as some botanists do, that other leafy plants may be intended. Some authorities believe that lettuce or endive could be among the plants of the scriptures. However, with the emphasis on the word leek in Numbers 11:5, coupled with onions and garlic, the similarities are apparent. Bulbous plants were so common among vegetables and flowers of the Holy Land, conditioned by that particular two-season climate of rainy and dry seasons, leeks are most likely the leeks we know today. Perhaps an even more compelling reason to identify the leek as the plant mentioned in the scriptures is the simple fact that it is still widely used today among the Jewish people, both in Israel and elsewhere around the world.

You can enjoy the sweet, subtle flavor of leeks wherever you garden. They are easily grown from seeds but do require a long period of time to mature properly, up to 130 days from the time you sow the seed. Plant seeds indoors for the earliest crop, for later transplanting outside, or sow outdoors in early spring when soil is warm if you have a sufficiently long growing season. Thin plants to stand 4 to 6 inches apart in rows about 2 feet apart. Leeks resemble huge scallions as they grow. The long thick necks may be blanched to improve their flavor. Simply hoe earth up around the base of growing plants and the necks will blanch naturally.

Harvest your leeks in the fall. You may use them in a variety of ways, especially in gourmet French cooking recipes. Since leeks are rather hardy, you may leave mature plants in the ground throughout the winter in most areas and dig as needed. For the coldest areas, it is best to dig and store your leeks in a cool basement area.

Leeks can, of course, be grown in pots and tubs. Their characteristic hollow leaves provide greenery on a porch or patio. You may even clip off leaves for use in seasoning soups and salads if you wish during the growing season.

GARLIC

Garlic, another bulbous relative of leeks and onions, also was deeply rooted in the soil of Egypt and the land of the Bible. This exceedingly pungent and strongly scented bulb remains popular through the countries of the Mediterranean today, as a seasoning primarily, but it is also eaten raw on slices of bread.

There are more than 60 different kinds of onion and garlic known to exist from the land of the Bible, so it is understandable that garlic should be identified in the scriptures. In the Hebrew Talmud, you will find that many kinds of food are to be regularly seasoned with garlic. It remains a favorite flavoring with Jewish people today.

Allium sativum is the Latin botanical name for garlic. This plant is a bulbous perennial with long, narrow, flat, and keeled leaves. The bulb is composed of bulblets, commonly called cloves, and these are the only part which are eaten or used for flavoring. A bulb may have 8 to 12 of these cloves.

Because of its popularity today to flavor such dishes as shrimp scampi and spaghetti sauce, tens of thousands of pounds of garlic are grown commercially in the United States, primarily in California, Louisiana, and Texas. As with many garden plants, you can grow it easily and can savor the best this plant has to offer when you have picked it fresh from your garden.

To grow garlic, follow the cultural techniques for improving soil as you would for onions. These plants also prefer clean cultivation, so pay attention to weekly weeding to eliminate any competition from unwanted plants around your garlic bed. You can grow a few in pots if you prefer or don't have outdoor growing room.

You may order garlic sets, which are actually the bulbs, from mail order firms, or obtain them from some garden centers. Separate the cloves before planting. A pound of garlic sets will plant about a 20-foot row. A few cloves will provide sufficient sets for several pots or planting in a windowbox with other herbs for your kitchen cookery needs.

LETTUCE

There are, according to learned botanists, potentially other vegetables from the scriptures which perhaps belong here, among them the castor bean, *Ricinus communis*. Since seeds are not readily available, it has not been included in this book.

The scriptures make reference to bitter herbs. Here again, botanical scholars have searched the land as well as researched ancient texts to accurately identify these bitter herbs. Those which are most clearly determined to be herbs, such as dill and sage, are included in the Herb section of this book.

However, among Biblical authorities who have explored the derivation of plants of the Bible, there seems to be a consensus that several other plants could likely have been intended in the reference to bitter herbs. These are lettuce, endive, chicory, and dandelion. Since these are more widely considered salad vegetables, they are included here.

The leaves of garden lettuce, *Lactuca sativa*, are often bitter when unbleached, as you may have noticed with some head lettuce you have grown previously. Of course, modern plant breeders have improved varieties of lettuce, which has to a large degree eliminated the former bitterness of earlier types. Endive, *Cichorium endivia*, also can have a somewhat bitter taste at times. In Greek translations of the Bible, you will find the word endive used in place of bitter herbs. That may be one of those mistranslations that has slipped into common usage as the Bible has undergone variations through translations through the years.

Other botanists believe that the term endive should be more loosely applied and actually covers several green herbs, including the young, tender leaves of the common chicory, *Cichorium intybus*. Today, in Egypt, as in days of old, various green herbs are still eaten. It is probable that the children of Israel during their years of bondage in Egypt did accept the custom of eating such leafy herbs with their bread and meat. Even the common dandelion, *Taraxacum officinale*, fits into that picture rather well.

If you wish to consider these "bitter herbs," as many authorities do, as represented by lettuce, endive, chicory, and dandelion, you may, of course. Here are tips for growing them, as vegetables, in your home garden. All respond to good cultural practices similar to those you would use for growing lettuce.

Lettuce is a universal favorite. Today it stands as one of the top ten most popular vegetables in America. Assuming that lettuce and these other salad crops are among the bitter herbs, plant breeders have achieved marvelous improvements for our growing pleasure. Today, these plants are not really bitter; rather, they are excellent to add new tang to salads.

Stately date palms rise in groves marking underground water sites in the Holy Land. (Photo courtesy of Neot Kedumim, Ltd., Kiryat Ono, Israel)

Figs are still a favorite fruit around Galilee. (Photo courtesy of Neot Kedumim, Ltd., Kiryat Ono, Israel)

Pomegranates shine brightly in the bright sun of the Holy Land. (Photo courtesy of Neot Kedumim, Ltd., Kiryat Ono, Israel)

A blossoming almond tree displays its glory in the "Garden of Choice Products" of Neot Kedumim in Israel. (Photo courtesy of Neot Kedumim, Ltd., Kiryat Ono, Israel)

Mature myrtle bushes line a wall at the top of the "Hill of the Menorah" in the Biblical Gardens of Neot Kedumim near Tel Aviv in Israel. (Photo courtesy of Neot Kedumim, Ltd., Kiryat Ono, Israel)

Wild flowers abound in the moist season in the Land of the Bible. (Photo courtesy of Neot Kedumim, Ltd., Kiryat Ono, Israel)

Masses of Yellow Flags grace the Land of the Bible in moist areas along waterways.
(Photo courtesy of Neot Kedumim, Ltd., Kiryat Ono, Israel)

Today glorious tulips in yellow and red can be traced to the Land of the Bible.
(Photo courtesy of White Flower Farm, Litchfield, Connecticut)

Today, lettuce is among the top ten vegetable favorites across America. You can grow many types in your gardens, including the Butterhead lettuce shown here, as you consider whether these are truly the "bitter herbs" of the scriptures. (Photo courtesy of National Garden Bureau, Inc., Los Altos, California)

Lettuce is available in a wide range of types and varieties. You can keep them growing outdoors from early spring right up to frost. Some varieties respond to indoor culture in pots and planters, too.

Lettuce actually prefers cool weather for best growth. In hot periods, lettuce tends to send up seed stalks and becomes extremely bitter. Since most of us prefer the more tender, less bitter leaves of lettuce, try these tips. To keep lettuce growing during summer's heat, try new loosehead and bibb-type lettuce plants. Some of these varieties trace their lineage back more than twenty-five centuries to the royal gardens of Persia and beyond. Plant breeders seemingly have incorporated heat-resistant ability in these new hybrids.

For best results, cultivate your outdoor soil 6 to 8 inches deep. Build humus content year by year, adding compost, manure, and mulch of old leaves, which can be turned or tilled under to improve the soil. Lettuce seedlings are sensitive to cold weather. To get a jump on spring, plan to sow lettuce seeds in flats, peat pots, or pellets on your windowsills or under lights in your basement. When danger of frost is past, you may transplant lettuce seedlings directly into the warm spring soil of your garden.

Because lettuce matures rapidly and is such tasty eating in salads and on sandwiches, it pays to make successive sowings. By successive planting, you will have a ready, steady supply of lettuce, as well as endive and dandelions if you wish, throughout the growing year.

Lettuce, being a leafy vegetable, needs lots of water, to grow quickly and lushly. Clean cultivation is best, but you can mulch if you wish. Be certain to water regularly, especially during dry periods. Plan ahead as you garden, sowing seeds of hot-weather-resistant types of loosehead and bibb lettuce directly outdoors for summer crops. By summer, you can again plant seeds of the cool weather varieties that will mature in the cooler days of fall. Information about the needs of these different types of lettuce is on the seed packs and is detailed in mail order seed catalogs as well.

Among crisphead or heading-type lettuce, which is the most popular, Great Lakes is excellent. It thrives under hot and adverse conditions, providing well-folded heads with tasty, brittle texture. Iceberg can stand some hot weather too, but both of these are best planted early, or for late season crops.

Butterhead-type lettuce is not usually grown commercially because this type lettuce doesn't harvest or ship well. For home gardens, the tender leaves and succulent flavor make this type superb. Buttercrunch is smaller, but its heavy, compact heads compensate for its size. Summer Bibb and Fordhook are other fine varieties.

Among loosehead lettuce, Greenhart matures quickly, in 45 days from seed. Black-seeded Simpson is early, crisp, and delicately flavored. Prizehead is crisp with excellent flavor.

Lettuce likes to eat well, so be sure to apply up to 4 pounds of 5-10-10 or similar fertilizer per 100 square feet of garden plot to insure satisfactory crops.

Dandelions may not appeal to you, but for tangy taste treats they are surprising and welcome additions to a salad. Some people who have their roots in Mediterranean cultures actually collect wild dandelion greens for boiling as well as for salads. Those greens can be bitter indeed. However, the domesticated dandelions with dark-green leaves can readily be grown from seeds you can obtain from several leading mail order firms.

Cultivated types produce stout hearts which may be eaten raw if the leaves are tied together to blanch the hearts. You can also use them chopped in salads, or boil the greens for a uniquely different vegetable.

Gourmets who favor light diets have helped make chicory popular again. This herb, or vegetable if you prefer to call it that, is well worth tending. Follow the same growing methods you would for lettuce and this plant will reward you well.

Sugarhat variety has elongated oval heads, which resemble Cos or Romaine lettuce. Outer leaves are medium green, but the inner leaves blanch from light cream to bright yellow, even without tying the outer leaves around the heads. The leaves are tender with a sweet, yet slightly tangy taste. You may use these leaves in salads or as cooked greens.

ENDIVE

Witloof, also known as French endive, produces tall, leafy plants with healthy root systems by fall. The headlike clusters of blanched leaves are 5 to 6 inches long, compact and tasty. This variety of chicory can be transplanted into boxes filled with sand and peat moss, and kept growing in a warm, dark indoor spot during the winter. You may also force Witloof chicory in pots or tubs to produce crisp, creamy yellow, blanched "chicons" which are great for salads. You'll need a fairly rich soil mixture of equal parts humus or loam, sand and peat moss. Plant seeds as you would outdoors, keep plants well watered and you'll be able to pluck these bittersweet leaves for much of the year as an indoor plant.

Endive provides slightly pungent but tasty salad greens in spring from early planting, or as a fall and winter crop from midsummer sowing. It stands well in the garden, but hot weather contributes to bitterness. Seed packets have detailed sowing and growing directions, which are similar to lettuce culture. Endive plants can be moved after frost to a root cellar or cool basement. Tie the outer leaves around the heads to blanch them.

Two varieties seem best, based on growing tests. Broad-leaved Batavian, or Escarole, has large, slightly twisted lettuce-like leaves. These form around plants up to 16 inches across. The heart is very deep, well blanched, and creamy white, with an almost buttery texture. Green Curled, also called Giant Fringed Oyster, makes an excellent crisp, tender salad green. It is finely cut with laciniated leaves that blanch at the center to creamy white. It too is delightfully tasty for your gourmet salads.

Vegetables of the Bible, from cool, sweet, refreshing melons to the tangy subtleties of these leafy herb plants, all respond to careful cultivation. As you grow and eat them, you too will be sharing some of the foods and experiences that are so deeply rooted in the land of the Bible.

We must remember that herbs and spices were highly valued in ancient days for many purposes. Today, we have refrigeration and have perfected other satisfactory ways to preserve food from one growing season to the next. In Biblical times, the people had no such methods. In order to preserve the foods which they grew, they relied on drying, which was a most common practice in the hot sun of the area.

Salting and smoking are also thought by experts to have been used to preserve food for those months when the land was dry, hot, and nonyielding.

Spices were one way to preserve food. Many authorities also point out that herbs and spices were undoubtedly utilized to cover the smell and flavor of meat and fish that was less than fresh. Even in the Middle Ages, without food-preserving methods, the people relied on other ways, including herbs and spices, to take their attention away from the taint of spoiled or nearly spoiled game and meat. It is logical, then, that herbs were utilized partially as a preservative as well as flavoring to cover up bad flavors in those ancient days.

Since spices and herbs were valued, they also served as important ingredients in trade, often in place of money. In St. Matthew 23:23 we read, "Woe unto you, scribes and Pharisees, hypocrites! for ye pay tithe of mint and anise and cummin, and have omitted the weightier matters of the law, judgment, mercy, and faith: these ought ye to have done, and not to leave the other undone."

If you wish to pursue other references to herbs, Cruden's Concordance lists them all, from Genesis through the Old and New Testaments. There are passages in Exodus and Deuteronomy, in Kings, Isaiah, Job, and Psalms, in Numbers, Proverbs, and Jeremiah.

You can muse on the meanings of the herbs in passages from St. Matthew, St. Mark, St. Luke, the books of Romans and Hebrews. Many thousands of students of the Bible have studied these passages and mused before you.

For this book, the best thinking of scholars and botanists alike has been considered in selecting the herbs of the Bible which should be included. Again, as with other plants from the scriptures, you will find only those which realistically can be grown successfully and which are generally available as seeds or plants.

One of the earliest mentions of a specific herb is in Exodus 16:31: "And the house of Israel called the name thereof Manna: and it was like coriander seed, white; and the taste of it was like wafers made with honey."

Manna is again likened to coriander in Numbers 11:7: "And the manna was as coriander seed, and the colour thereof as the colour of bdellium."

With the renewed popularity of gardening across America in the past near-decade, many people have rediscovered the pleasure of growing herbs. With that rediscovery, which is continuing in many households today, has come the realization that herbs can add new zest and spice to life. Best of all, many of these herbs are easy to grow, are hardy, and don't really require much care. In fact, because of their aromatic nature, even insect problems are reduced.

Many herbs bear lovely blossoms, but few of us wait to see the flowering display. Instead, we grow our herbs, as you can, for their leaves, stems, stalks, and seeds. Actually, the best part of herbs is their flavor and flavoring ability. Even a pinch of one herb or a leaf of another can make meals come alive and awaken dulled appetites. You can sprinkle herbs on salads or add them to soup. You can blend them with vegetables or into gravy and sauces and even cook them to season meat.

Another compelling point in favor of herb gardening is that you can enjoy them year round. Once your herbs mature, you can snip their appropriate parts, dry them, and store them for future use.

Some herbs like sun, which is understandable, considering the climate in the land of the Bible. Others actually prefer more shaded areas. You'll find the appropriate requirements of each in this chapter.

You can grow herbs in an outdoor garden, in a site that has reasonably fertile soil. Herbs, being hardy, may surprise you with their ability to grow even in poorer soils. However, like many other plants, they respond best when well rooted in the best possible growing environment. Since in cooking with herbs you need a pinch or a piece only, it saves time to plant your herbs of the Bible near the kitchen door. You can grow herbs in their own bed, as a border with flowers, in a rock garden or perhaps use an old-fashioned wagon wheel to neatly divide the growing spots for the different herbs.

If you prefer to grow some indoors, start with this potting mix. Combine equal parts of sandy loam or sterilized potting soil with sand and peat moss. You may substitute composted humus for the peat. Any type container is satisfactory, so long as it has provisions for drainage. Herbs, like most plants, cannot tolerate soggy roots.

Sowing seeds of herbs is identical whether you garden in containers or outdoors. The most important ingredient is patience. Herb seeds can be notoriously slow to germinate.

For your purposes, some herbs may be more practical to grow for their use in your cooking. Don't overlook the value of others, especially as unusual specimen plants in rock gardens, in beds and borders along a path or wall. Their significance as plants of the Bible has its own merit too.

A L O E

"And there came also Nicodemus, which at the first came to Jesus by night, and brought a mixture of myrrh and aloes, about an hundred pound weight." St. John 19:39

There is, according to the most respected scholars, a difference between the aloes of the Old Testament and the aloes mentioned in the New Testament. We may tend to confuse the true identity even further by assuming that the American aloe, *Agave americana*, is involved somehow. In reality, the common American aloe, more properly called the century plant, is not related to the aloes of the Bible.

The aloes of the Old Testament, studying the content in which they are mentioned, are most likely trees. Reviewing the Greek translations and Hebrew texts, some authorities see the eaglewood tree, *Aquilaria agallocha*, as the Old Testament aloe. The context in the Douay, Moffatt, and Goodspeed translations seems to indicate a large tree and, in fact, the Moffatt translation does substitute the word oaks for aloes.

Honest scholars have seemed to translate the Bible through the centuries without the necessary knowledge of botany and plant physiology. Where some believe the aloe to be eaglewood trees or even sandalwood, *Santalum album*, neither of these trees was native to the Holy Land.

Since the debate about the true identification of the aloes mentioned in the Old Testament rages, you may wish to refer to the passages in the Bible. In Psalms 45:8, you will read: "All thy garments smell of myrrh, and aloes, and cassia," and in Proverbs 7:17: "I have perfumed my bed with myrrh, aloes, and cinnamon." In both these references, it appears that, grouped with other perfumes, scents, and spices, the aloes may indeed be a type tree that yields resinous gum from which incense is made, though which one, it seems impossible to say for certain.

Therefore, the aloes of the New Testament mentioned by St. John in 19:39 deserve closer attention. It is known that the ancient Egyptians had perfected the art of embalming to a high state. The juice of the true aloe, *Aloe succotrina*, was known to the Egyptians, and it is most likely that this drug, used in highly perfected embalming art, was the aloe substance brought by Nicodemus to anoint the body of Jesus.

The *Aloe succotrina* is a succulent plant with thick, stiff, fleshy leaves, as are all succulents. Its smell is disagreeable and the taste bitter, botanists point out. Although the plant is a native of the desert areas of east-coast Africa, it is strikingly similar to other aloes that can be grown as houseplants. Since it is nearly impossible to obtain the exact species to duplicate every plant of the Bible, you can perhaps be satisfied with close substitutions.

In their native habitat, some aloes become truly treelike in their growth. Few homes can accommodate such huge specimens. However, by selecting close relatives, you at least can have them as family representatives. Since most aloes are sun lovers, the best place to grow them is outdoors on the patio or balcony in summertime. Indoors, sunny windows are the best locations.

Actually, you have a wide choice of aloes to grow, from the Spiny Aloe, *A. africana*, to the Tree Aloe, *A. arborescens*. This treelike plant with branching stems carries spreading rosettes of swordlike tapering leaves. In winter sun it often blooms with vermilion to yellow flowers in long, spiky clusters. Since it does bear some resemblance to the ancient aloe, this may prove pleasing. It is especially easy to grow in sandy soil mix and can be propagated by top cuttings or suckers.

Aloe vera chinensis is an Indian medicine aloe, a small Asiatic form with fleshy, lanceolate leaves that curve at the tips. The leaves are blue-green in color with white markings. This and the aloe vera or Medicine plant aloe are easily raised on your windowsill.

The *Aloe vera* is a short-stemmed, freely suckering plant with dagger-shaped and channeled leaves of bluish or grayish green in a rosette pattern. Since ancient times the juicy pulp of this variety has been used to make the bitter-aloe thickened sap which is still used as a poultice to heal burns and cuts in its native areas of Africa.

If you aren't especially fussy about which aloe you wish to have represent the aloe of the scriptures, you can choose from dozens of cultivars which have been developed by nurseries and species propagated from native plants as well.

The keys to success in cultivating aloes are providing them with ample sun and watering sparingly at least once monthly. A gritty soil mixture is desired. Best propagation is by dividing or planting cuttings or the suckers which may form.

CORIANDER

"And the manna was as coriander seed, and the colour thereof as the colour of bdellium." Numbers 11:7

From its mention in the Bible, and tracing its rightful place through the ages, we realize that coriander has been used as a food for flavoring in soups and salads. It also has been a favored ingredient in hot curries and sauces. To the ancients, it was used also as medicine, as an aromatic and carminative.

Coriander fruit, improperly called seed, comes from the *Coriandrum sativum* plant, which has been determined to be a native of Asia Minor and Europe. The plant produces a slender, erect, hollow stem that rises one to two feet tall. It has bipinnate leaves and small flowers of pink to white. The fruit itself is globular and smooth with five indistinct ridges. Since coriander thrives in most garden soils, you can grow it with other herbs that prefer good sun. It is an annual, so unless you grow it indoors, you must resow it every year. Today, home gardeners are using coriander seeds as coatings for confections and ground up for flavoring in cookies, cakes, and bread.

Coriander differs from dill and caraway in its globular, pearl-like seeds, which are quite aromatic. The leaves are also aromatic and are used in soups, and to flavor puddings and wines. Today, coriander is used as a spice in many Arab countries.

To grow it well, follow the general directions for herbs, and more specifically those for dill.

DILL

"Woe unto you, scribes and Pharisees, hypocrites! for ye pay tithe of mint and anise and cummin, and have omitted the weightier matters of the law. . . ." Matthew 23:23

Reading this passage of the New Testament, you might be compelled to purchase anise seeds to plant in your Biblical garden. Many people have made this same error, which seems to have been perpetuated for several centuries in a translation of the Authorized Version of the Bible. Although the Douay and O'Hara versions of the Bible also retain the word anise, almost all authorities, Biblical scholars and botanists alike, now agree that the true anise as we know it botanically, *Pimpinella anisum*, is not the anise mentioned in the scriptures.

Considering the roots of the herb dill, *Anethum graveolens*, and comparing plant anatomy with other factors, dill is most likely the herb mentioned in the passage from St. Matthew. Dill has been widely cultivated for centuries for its seeds, which are aromatic and carminative, similar to those of caraway seeds. The leaves also have been used in preserving cucumbers as pickles through the years. Among many earlier civilizations and present ones, dill has been widely used for flavoring as well as in medicine. It can be found growing wild in Palestine to this day, especially on the plain of Sharon. We opt with the majority of experts for the idea that this plant called anise is in reality dill.

Dill is an annual which sets its seeds in late summer and early fall. It is a tall plant with finely divided leaves and flowers. The feathered, light-green leaves add an attractive, almost fernlike look to your garden land-scapes. Both the leaves and seeds of dill are useful, but you must be watchful to harvest the small brown seeds before they drop in late summer or fall. Dill flowers that produce the seeds are a greenish yellow and form an umbrella-shaped cluster, from 3 to 6 inches across.

In outdoor gardens, dill may grow 3 to 4 feet tall from its deep tap-root. You can plant seed in early spring when danger of frost has passed. Place 3 to 5 seeds per inch, about one quarter inch deep, in sunny loca-tions. When sprouts are a few inches high, thin seedlings to 5 or 6 inches apart.

Since dill does have a long taproot, you will need a deep pot if you wish to try growing dill indoors. Fill a large pot with fertile soil, plant your seeds and keep the pot well watered in a sunny location. Dill is slow-growing at first, so be patient. Keep the soil moist, and eventually you'll see results.

To harvest dill, cut the best leaves and place them in a warm, dry area. Leave them for several days. After dill is thoroughly dry, crumble it and place it in airtight glass containers. To harvest seeds, shake the heads over a white sheet to release the seeds. They're easier to see this way. Rub them with your fingers to remove the chaff and then store seeds in their own glass container after they are fully dried.

You can cut plants with leaves and seed heads and hang them in a cool, dry place to air dry. But be sure to place a clean sheet beneath them to collect the seeds which fall.

FRANKINCENSE

"And the Lord said unto Moses, Take unto thee sweet spices, stacte, and onycha, and galbanum; these sweet spices with pure frankincense: of each shall there be a like weight." Exodus 30:34

"And thou shalt put pure frankincense upon each row, that it may be on the bread for a memorial, even an offering made by fire unto the Lord." Leviticus 24:7

Burning incense has been a traditional ritual for many different religions through the years. As you read the Bible more carefully, you will find literally dozens of references to the use of incense. Frankincense was highly prized among the scents by the ancients in the Holy Land. In fact, along with myrrh, it appears in one of the most often quoted passages in the Bible.

The story of the Magi, the Wise Men who came to pay homage to the Christ Child, is retold during the Christmas and Epiphany seasons every year. "And when they were come into the house, they saw the young child with Mary his mother, and fell down, and worshipped him: and when they had opened their treasures, they presented unto him gifts; gold, and frankincense, and myrrh" (St. Matthew 2:11).

Both frankincense and myrrh were probably the most favored among ceremonial plants for the valuable gums which the ancients derived from them. Judging from the choice by the Magi to bring frankincense and myrrh as gifts to the Christ Child, these incenses were indeed in the same prized category as gold.

Although this tree also is seldom available for you to grow, as a plant of the Bible it does belong in this book for its historic significance. Most authorities agree that *Boswellia carterii* is the true frankincense plant. Some feel that two other species can be considered, *Boswellia thurifera* and *B. papyrifera*. All provide a similar gum which does burn readily, spreading a pungent perfume, as the ancients did so frequently in their religious ceremonies. Frankincense can be literally translated as free-burning, as indeed it is.

The Frankincense trees, of the genus Boswellia, grow in southern Arabia, across Abyssinia, and along the east coast of Africa. Other spices are found in India and the East Indies. Recalling your school studies of ancient history, you may remember the caravans that traveled to the Orient in search of spices and incenses, which were valued articles of trade.

These trees, from which the incense is obtained, are of relatively large size and are in fact related to the terebinth tree and to those scrubbier trees that yield myrrh. The tree's flowers are star-shaped and white with some rose tinting. The leaves are compound and include 6 to 9 leaflets. In some respects, these trees look like a mountain ash, which is popular in many parts of the northern United States.

As you read through the scriptures to learn more about frankincense and its uses, you will find it mentioned twenty-two different times. Most often it is referred to for its use in religious worship, and for its worth in bestowing honor or tribute.

In Kings and Chronicles, Nehemiah and The Song of Solomon through Isaiah and Jeremiah, you will find references to many ceremonies in which incense was burned.

Frankincense is an age-old incense which is used for other purposes as well even today. It is obtained from the sap of plants.

The incense itself is still gathered in much the same way it was during the time of Christ. Successive incisions are made in the bark of the trunk and branches of living trees, much as rubber trees are tapped for their latex in other parts of the world. The first incision is especially valued since it produces the purest resinous gum, which is almost white. The gum becomes yellowed as the tapping continues. Eventually this resin dries into a semitransparent, yellow color, which has a bitter taste. When burned as incense, it gives off a powerful balsam-like odor. In ancient days it was also used for fumigating and purifying, as well as burning in the temples.

Today, in the Middle East and parts of Asia, frankincense remains one of the most desired of the natural incenses. It is gathered in Africa and the Middle East as well as in India, from related species of trees. According to some botanists, it is as valued today in many parts of the world as it was in the time of the Bible.

The Biblical Garden planners at the Cathedral of St. John the Divine in New York City do intend to add frankincense trees to their plantings in the future. However, at present, obtaining any of these true species is nearly impossible. You may elect to grow the mountain ash, which is similar in appearance but not actually related. Or, you may feel that the closest you can get is the tree that produces the common European frank-incense. This is obtained as resin from slits in the bark of the Norway spruce fir, *Abies excelsa*, but again, it is not really related to the true frank-incense-producing trees of the Bible.

MARJORAM

"And he spake of trees, from the cedar tree that is in Lebanon even unto the hyssop that springeth out of the wall: he spake also of beasts, and of fowl, and of creeping things, and of fishes." I Kings 4:33

Hyssop is a bushy herb in common use among the Hebrews of ancient times and, to this day, travelers can find this herb sprouting out of the ancient walls in the Holy Land.

There remains some controversy about the true identity of the hyssop, and it is true that many other plants do "spring out" of holes and crevices of walls in Israel. However, botanists have focused their attention on this subject and many are agreed that the hyssop is most likely the Syrian marjoram, a popular herb today as it was in Biblical times.

In Psalms 51:7, David's confession of sin and humble prayer for forgiveness tells again of the hyssop: "Purge me with hyssop, and I shall be clean: wash me, and I shall be whiter than snow."

If you read again the passages of Numbers, hyssop appears in chapter 19:6: "And the priest shall take cedar wood, and hyssop, and scarlet, and cast it into the midst of the burning of the heifer." Later, in verse 18: "And a clean person shall take hyssop, and dip it in the water, and sprinkle it upon the tent, and upon all the vessels, and upon the persons that were there, and upon him that touched a bone, or one slain, or one dead, or a grave."

It would seem, thinking about these different references to hyssop, that it must indeed be an herb which is valued for its particular properties in the rituals and in the cleansing of the people.

Many writers and scholars have continued their debate about the word hyssop. Some argue that it is really the well-known garden herb now called hyssop, *Hyssopus officinalis*. Others argue in favor of the caper, *Capparis sicula*, which is a spiny shrub found in desert areas and the rocky parts of Israel. Still other authorities favor the sorghum, *Sorghum vulgare*, as the most likely plant.

Perhaps, as most believe, the word hyssop in different books of the Bible refers to several plants. For the hyssop of the Old Testament, most experts vote for the Syrian marjoram, *Origanum maru*. In fact, marjoram is most common among rocks and terrace walls, which fits the scriptural passage perfectly. We vote with the majority.

Marjorams are members of the mint family. Under favorable conditions they will mature to 3 feet tall. Restricted in rock crevices or poor soils, they are usually much smaller. Marjoram has erect, stiff, hairy branches with thick-textured leaves. It bears white flowers at blossoming time. If you examine the hairy-stemmed marjoram, it is logical that it would hold water to serve as a sprinkler in religious ceremonies, as in Exodus 12:22: "And ye shall take a bunch of hyssop, and dip it in the blood that is in the bason, and strike the lintel and the two side posts with the blood that is in the bason . . ."

From its roots in western Asia and the Mediterranean, marjoram has spread its sweet scent to many lands. Both Greeks and Romans prized the aroma of marjoram and used sweet marjoram in creating the crowns worn by bridal couples. As a flavoring, this herb has wider applications than most. It is a natural with lamb and fish, but also adds its stimulating bouquet to eggs, vegetables, soups, stews, and stuffing for poultry. Gourmet cooks point out that it is best to add marjoram to vegetables and soups just before the end of cooking time, so its delicate perfume is not lost in overcooking.

Marjoram is a tender, sweet-smelling herb that thrives as a perennial in southern areas, but it must be replanted as an annual each year in colder northern climates.

You can grow marjoram from seeds or cuttings in a sunny location. Plant 5 to 8 seeds lightly, only a quarter inch deep, in late spring when the soil is well warmed. Be patient; the seeds may require several weeks to begin sprouting. After 6 weeks, thin the seedlings to stand 6 to 8 inches apart. Indoors, use a fertile soil mixture containing humus and sprinkle the tiny seeds evenly on the surface. Keep moist until they sprout and then thin to leave the strongest two or three in the pot.

Marjoram is one of the few herbs that prefer sweet or alkaline soil rather than acid soil. Because of this, add a little lime around each plant in the ground during spring and fall, and to the soil of your container as well. Water weekly and tend carefully. To keep plants from spreading, prune them every few weeks and remove blossoms. These steps will keep the leaves sweeter.

You can use marjoram in salads, casseroles, or in making herbal tea, as well as seasoning for meats, soups, and stews. To save it, cut and hang branches in a dry, warm room that is well ventilated. When leaves are crisp, strip them from the stems and put them away, either whole or chopped, in airtight jars.

M Y R R H

"All thy garments smell of myrrh, and aloes, and cassia, out of the ivory palaces, whereby they have made thee glad." Psalms 45:8

"And when they were come into the house, they saw the young child with Mary his mother, and fell down, and worshipped him: and when they had opened their treasures, they presented unto him gifts; gold, and frankincense, and myrrh." St. Matthew 2:11

Among all the passages from the Bible, this from St. Matthew may be the one that is most widely quoted. As the Wise Men came to see the Holy Child, they brought with them gifts of gold, frankincense, and myrrh, their most valued treasures. What may not be clearly understood is that two of these treasures offered by the wise men are actually derived from plants.

Myrrh is a gum resin derived from the myrrh plant, *Commiphora myrrha*. True myrrh was highly valued and esteemed by the ancients both as a perfume and as incense in the temples. It was used as an unguent and also in embalming.

Some authorities believe that the word myrrh in various translations of the Bible may also refer to a related plant, *Commiphora kataf*. Both of these trees are native to the coast of eastern Africa, Abyssinia, and Arabia. From ancient records it is known that the gummy substance taken from these trees provided the commercial myrrh of antiquity. In fact, tracing through botanical texts and other references, it is obvious that myrrh was widely used throughout the Holy Land and its adjacent areas.

These two most likely species of myrrh are similar and forbidding in appearance. These are low, scrubby, stiff-branched, and thorny shrubs or small trees. Today they grow in rocky areas, especially in limestone hills in the Middle East and many parts of North Africa. The three-part leaves grow in clusters on the wood accompanied by sharp thorns.

Myrrh is among the valued treasures that the scriptures say the Wise Men brought to the Christ Child. Like frankincense, it is obtained from the sap of a shrub-like plant and is used as incense.

This strange plant, so deeply rooted in the scriptures, bears an oval, plumlike fruit. Its bark and wood are aromatic, but it is the resin, which is tapped from the bark and allowed to dry to a golden-brown color, that has won this rather odd and ugly plant its fame. Myrrh is sold as a spice in the Middle East and still is used there as a medicine today.

Thumbing through the scriptures, you can find allusions to the value and desirability of myrrh to the people of the Bible. In The Song of Solomon 1:13, you can read, "A bundle of myrrh is my well-beloved unto me . . ." and later in 3:6, "Who is this that cometh out of the wilderness like pillars of smoke, perfumed with myrrh and frankincense . . ."

In The Song of Solomon, myrrh again is prominent in 5:5 and 5:13: "I rose up to open to my beloved; and my hands dropped with myrrh, and my fingers with sweet smelling myrrh, upon the handles of the lock . . . His cheeks are as a bed of spices, as sweet flowers: his lips like lilies, dropping sweet smelling myrrh."

Myrrh was praised by David and Solomon and is described again in St. Matthew, St. Mark, and St. John, should you care to pursue the other mentions of myrrh in the scriptures.

If you wish to try growing the myrrh tree, you may be disappointed, since it is not generally available in America. However, as more Biblical gardens are cultivated, it may be possible to obtain rooted saplings. There are plans to plant myrrh trees at the Biblical Gardens of the Cathedral of St. John the Divine, according to its founder and designer.

Some students of the Bible have thought that the European pot herb, *Myrrhis odorato*, may have been the plant that Solomon grew in his gardens. This plant, however, is a member of the carrot family. Although pleasantly fragrant, it is most unlikely that this European plant would have been grown in the land of the Bible, most authorities agree.

SAGE

"And he made the candlestick of pure gold: of beaten work made he the candlestick; his shaft, and his branch, his bowls, his knops, and his flowers, were of the same: And six branches going out of the sides thereof; three branches of the candlestick out of the one side thereof, and three branches of the candlestick out of the other side thereof." Ex. 37:17–18

Through the ages, artists often have borrowed from nature as they created their works of art. The Egyptians, Greeks, Romans, like the Hebrews, have adorned buildings with replicas of plants and flowers.

According to Biblical scholars, this passage from Exodus can be traced to one popular herb that was common throughout the Holy Land. That plant is sage. The Judean sage, *Salvia judaica*, grows to 3 feet tall and can be found today in most of Israel. The stems are 4-angled and stiff with paired leaves. When pressed flat, sage is likened by Biblical scholars to the seven-branched candlestick which is the traditional Jewish symbol, the menorah. If you examine a sage plant with its central spike and three pairs of lateral branches, you will notice that each bends upward and inward in a symmetrical pattern. On the branches are whorls of buds, which perhaps gave the artist the knops on the Biblical golden candlestick.

Siding with the scholars, that the sage was and is a plant of the Bible, we also know it has proved useful as a flavoring as well as for medicinal purposes for centuries. Sage is a distinctive herb with a most pungent aroma. It is a hardy perennial which will grow 2 to 3 feet tall, bearing lavender to whitish flowers.

The oval leaves, which may be a few inches to several inches long, are grayish green in color and somewhat coarsely textured. There are, of course, other types of sage, such as the golden sage with yellow variations on leaf edges. The garden sage is the one to grow, however, to be closest to the true Biblical plant.

Plant several seeds about one quarter inch deep in early spring. Be patient as you water the seedbed regularly, because sage, like other herbs, may require 3 to 4 weeks to sprout. Thin seedlings 10 to 12 inches apart. Because sage begins its life slowly, keep weeds under control to eliminate any competition for moisture or nutrients.

Indoors, sage prefers a sunny window that faces south or east. Place 3 to 6 seeds in a 6-inch pot. Thin to the two strongest seedlings. Sage will reward you well without much care. In fact, overwatering can be harmful, indoors or outdoors. Mulching outdoors is helpful, but don't water unless soil is really dry. Normal rainfall is usually satisfactory.

You can harvest sage by picking the leaves periodically. To improve the quality of the leaves, prune the woody stems occasionally. This encourages new branches to form which will provide more tender young leaves.

To dry sage, cut sprays or bunches and hang them in a cool, dry spot. You can also strip the leaves from the stems and place them on a clean screen or sheet to dry in the sun.

Sage leaves are mainly employed in stuffings and sausage as well as meat flavoring. This pungent herb adds a piquant or sharp taste to vegetables, and many recipes call for its use with beans, in stews and soups that have their origin in Mediterranean lands.

When the leaves are crisp and brittle, crush them and store them in airtight glass jars for future cookery fun and flavor.

WORMWOOD

"Lest there should be among you man, or woman, or family, or tribe, whose heart turneth away this day from the Lord our God, to go and serve the gods of these nations; lest there should be among you a root that beareth gall and wormwood." Deuteronomy 29:18

Wormwood is another of the ancient herbs that retains to this day visions of bitterness and distaste. In fact, Webster's dictionary provides a definition of wormwood as "something bitter, galling, or grievous." Some plants seem cursed with a bad reputation through the ages.

As you read other passages from the Bible, you will find that wormwood does seem to be held in low regard. In Proverbs 5:4, "But her end is bitter as wormwood, sharp as a two-edged sword."

The theme of bitterness and the ill omens associated with wormwood are even more clearly indicated in Lamentations 3:15, "He hath filled me with bitterness, he hath made me drunken with wormwood," and later in verse 19, "Remembering mine affliction and my misery, the wormwood and the gall."

In Jeremiah 9:15: "Therefore thus saith the Lord of hosts, the God of Israel; Behold, I will feed them, even this people, with wormwood, and give them water of gall to drink."

You may find other references to wormwood in Amos and in Revelation too. It would seem, reading through these passages and contemplating their meanings, that wormwood was far from being a prized herb of Biblical days.

There are many species of wormwood, which include annual, biennial, and even perennial plants, often woody plants with strong, aromatic odor. The sagebrush of our own Wild West belongs to the same family as the wormwood of the scriptures. Biblical scholars and botanists, in their tracing of the identity of wormwood, believe it is most likely *Artemisia herba alba* or *Artemisia judaica*.

One species of wormwood, despite its reputation and association with bitterness, has been used in making a drink called absinthe. Perhaps in olden days it also was used in such libations, leading to the references "drunken with wormwood" that you find in Lamentations. Wormwood, whenever mentioned in the New or Old Testament, always serves as a symbol of bitterness. Although there are several species, all do secrete a bitter juice which serves a wide variety of purposes from the distilling industry to medicinal purposes.

You may grow wormwood if you have a well-drained, sunny area. The plants grow 1 to 3 feet tall, have small leaves, somewhat hairy stems, and yellow flower heads at the end of the branches.

Since seed of wormwood is not readily available, and since most gardeners seem uninterested in cultivating a plant with such a dour reputation, we have elected to concentrate on those with more appeal than wormwood.

CHAPTER IX

FRUITS
OF THE BIBLE

"And God said, Let the earth bring forth grass, the herb yielding seed, and the fruit tree yielding fruit after his kind, whose seed is in itself, upon the earth: and it was so." Genesis 1:11

"And God said, Behold, I have given you every herb bearing seed, which is upon the face of all the earth, and every tree, in the which is the fruit of a tree yielding seed; to you it shall be for meat." Genesis 1:29

"And they came unto the brook of Eshcol, and cut down from thence a branch with one cluster of grapes, and they bare it between two upon a staff; and they brought of the pomegranates, and of the figs." Numbers 13:23

"And they told him, and said, We came unto the land whither thou sentest us, and surely it floweth with milk and honey; and this is the fruit of it." Numbers 13:27

From the beginning of the Bible in Genesis through its many poetic and prophetic books, fruit trees are glowingly described and praised, as well they should be. Fruits were one of the most important foods for the children of Israel and nourished them faithfully in the days of the Bible.

Figs and dates, palms, olives and pomegranates, vines laden with grapes to eat and for making wine are vividly and colorfully described throughout the scriptures. As you reread the holy words, you'll find more than seventy references to the "fruitful vine" and grapes. No doubt these plants were most prized by the people who tended their vineyards throughout Biblical history. Not only did the vines yield precious grapes, but the crops were dried into raisins for use when the harvest was over, and they were squeezed into wine that would last until the harvest of grapes began the following year.

Olives also are widely mentioned in almost as many references as are the fruitful vines. They too had a prominent place in the life of the people. Olives provided their tasty fruit as well as their oil, which was a household necessity. Olive oil substituted for butter and fat. It also supplied the holy oil for anointing during ceremonial occasions and for lighting the lamps of the temple as well as people's homes.

Olives have a special significance beyond their value as food for the people. As you read in Judges 9:8, "The trees went forth on a time to anoint a king over them; and they said unto the olive tree, Reign thou over us." During the time of Noah when the waters had receded after the terrible flood, you find this often-quoted passage from Genesis 8:11: "And the dove came in to him in the evening; and, lo, in her mouth was an olive leaf pluckt off: so Noah knew that the waters were abated from off the earth."

From the writing of the scriptures to this day, the olive branch has been a symbol, along with the dove, of peace and promise of better times to come. As you explore the scriptures further, in the many dozens of passages about olive trees and their oil, you will realize how vital this plant was to the people of the Bible.

In Leviticus 2:1–2, the oil of the olive tree is prescribed for use in important ceremonial occasions and is emphasized again and again: "And when any will offer a meat offering unto the Lord, his offering shall be of fine flour; and he shall pour oil upon it, and put frankincense thereon: And he shall bring it to Aaron's sons the priests: and he shall take thereout his handful of the flour thereof, and of the oil thereof, with all the frankincense thereof; and the priest shall burn the memorial of it upon the altar, to be an offering made by fire, of a sweet savour unto the Lord." Further in verses 4 through 6 you will find this reference: "And if thou bring an oblation of a meat offering baken in the oven, it shall be unleavened cakes of fine flour mingled with oil, or unleavened wafers anointed with oil. And if thy oblation be a meat offering baken in a pan, it shall be of fine flour unleavened, mingled with oil. Thou shalt part it in pieces, and pour oil thereon: it is a meat offering."

As you read the many other passages, it is apparent that olive trees were indeed one of the most important of all trees in the Holy Land. Biblical gardens, it must be remembered, were not necessarily gardens as we think of them today. Rather, they were orchards of figs, olive trees, and palms where the people could retire to find shade and quiet during the heat of the day. The Gardens of Gethsemane are thought to have been such orchards of productive fruitfulness.

Fig trees too played their necessary part in feeding the people of the land of the Bible. This nourishing fruit is borne by these trees today in

Israel as it is in many parts of the Middle East and other Mediterranean countries, both in commercial orchards and in private home gardens.

In Deuteronomy 8:7–8, Moses reminds the children of Israel of God's goodness to them: "For the Lord thy God bringeth thee into a good land, a land of brooks of water, of fountains and depths that spring out of valleys and hills; A land of wheat, and barley, and vines, and fig trees, and pomegranates; a land of oil olive, and honey."

In Proverbs 27:18, figs again are prominent: "Whoso keepeth the fig tree shall eat the fruit thereof: so he that waiteth on his master shall be honoured."

In I Kings 4:25, you will find stories of Solomon's kingdom with its orchards: "And Judah and Israel dwelt safely, every man under his vine and under his fig tree, from Dan even to Beer-sheba, all the days of Solomon." Those must have been most productive and peaceful times for the children of Israel.

In Old and New Testament alike, the fig tree and its vital fruits are emphasized. Considering the reliance that the people of the Holy Land placed on their orchards, it is understandable that their fruit trees and vineyards figure so dramatically in the scriptures. Often the references were allegorical, meant to teach lessons for the people, using the trees as symbols in the process. At other times, the eloquent passages of other books of the Bible laud the trees for their beauty as well.

Studying the scriptures and reviewing botanical texts, it is clear that the olive, fig, and date are natives to the Holy Land and its neighboring areas. Although some of the species may have been brought by wandering tribes and early caravans from the deserts of North Africa and Arabia, their roots have long been established in the land of the Bible.

The date from the date palm is truly the staff of life in desert areas. Not only does it provide rich fruits to feed the traveler, but it is a sure sign that water lies beneath the seemingly dry desert sand. In the deserts of the Sinai and the Negev, date palms rise today as they have for aeons, providing their sweet goodness for man and beast.

During the intervening years, the land of the Bible also saw times of privation. Insects and drought were two natural phenomena that plagued the people of the countryside. Through the centuries, warring tribes and armies from other nations also pillaged the Holy Land. Fields were destroyed, orchards cut and burned, forests leveled. Man in his continuing inhumanity to man greatly altered the face of the land and the course of history.

But still the trees and the plants of the Holy Land endured. There were times of adversity, cruel and hungry times, as the scriptures so clearly state in Judges 15:5: "And when he had set the brands on fire, he let them go into the standing corn of the Philistines, and burnt up both the shocks, and also the standing corn, with the vineyards and olives."

There were understandably hard times when plagues of locusts devoured the crops and denuded the trees, as explained in Joel 1:7: "He hath laid my vine waste, and barked my fig tree: he hath made it clean bare, and cast it away; the branches thereof are made white," and further in Joel 1:10–12: "The field is wasted, the land mourneth; for the corn is wasted: the new wine is dried up, the oil languisheth . . . The vine is dried up, and the fig tree languisheth; the pomegranate tree, the palm tree also, and the apple tree, even all the trees of the field, are withered: because joy is withered away from the sons of men."

In recent years, plagues of locusts and caterpillars have periodically invaded the lands of the Bible. Just a few years ago, in fact, army worms, a form of caterpillar, marched by the billions across Egypt, laying waste the farms and orchards there.

Throughout recorded history, we have read and seen how natural phenomena like insects and diseases have ruined crops and orchards. Likewise, anyone who has been involved in any war has seen the ravages war leaves on the plants in the path of raging armies. One rule of warfare, it seems, is to destroy the food supply so that an enemy cannot obtain provisions from the land. That too was the way of life during Biblical days and through the centuries that have passed since.

Fortunately, plants have greater resilience than we expect. Even after insect plagues and forest fires, seeds that remain sprout again. Trees that have set their roots deeply in their native soils somehow find inner strength to send up new shoots and grow again. Men and women, dedicated to restoring the ravaged land, also have been a vital factor in rebuilding the land's productivity. Today, in modern Israel, great acreage has been reclaimed and made to bloom and bear abundantly again. Modern irrigation, improved fertilizing, the use of new hybrid plants with their hardiness and increased vigor, have made the deserts blossom anew.

In this chapter, you'll find many references to the fruits of the scriptures. Because there are so many passages that mention these trees and their crops and vineyards with their grapes and wine, we cannot possibly provide all the references. If you wish to pursue them all, Cruden's Concordance is a likely source to guide your reading in the scriptures. This book is intended more as a guide to growing, so that you too can find the most appropriate passages from the Bible and learn how to grow the plants that you find mentioned there.

Since it is impractical for many home gardeners to grow all the tree fruits identical to those from the Holy Land, you'll also find the best and closest representatives from among the families of these plants. If, for example, you live in an apartment or a northern area, it is impossible to enjoy the pleasures of large fig trees and date palms. However, members of the *Ficus* and *Phoenix* families can be grown as indoor plants. At least with them, you can have the feeling of the scriptures with those similar plants growing where you live.

APPLES AND APRICOTS

"And out of the ground made the Lord God to grow every tree that is pleasant to the sight, and good for food; the tree of life also in the midst of the garden, and the tree of knowledge of good and evil." Genesis 2:9

"But of the tree of knowledge of good and evil, thou shalt not eat of it: for in the day that thou eatest thereof thou shalt surely die." Genesis 2:17

All of us who have read the Bible are familiar with this reference to the "tree of life," "the tree of knowledge of good and evil." It is the apple, or so it has been assumed to be for as long as the scriptures have been read. Ever since translations have made the scriptures available to masses of people, the apple has carried this stigma with it wherever it has grown, and this stigma is, most likely, totally unjustified.

The apple of the Bible is another of those puzzling problems of correct botanical identification. From the earliest translations, discussion, debate, and heated argument have revolved around this one word and this tree. Reading older scriptures and earlier translations, scholars have determined that the "apple" tree was a tree which provided shade and its fruits yielded fragrance, sweet taste, and attractive sights to behold.

Scholars through the ages have argued that the common apple, *Malus pumila*, is the apple of the scriptures. Some modern authorities still persist in this belief, pointing to apple orchards that produce large and delicious fruits.

Other references through the Bible mention apples and their comforting appeal. In The Song of Solomon 2:3, you can read about the virtues of this fruit: "As the apple tree among the trees of the wood, so is my beloved among the sons. I sat down under his shadow with great delight, and his fruit was sweet to my taste."

Further passages from The Song of Solomon return to the pleasures of the "apple" tree. In 2:5 you find: "Stay me with flagons, comfort me with apples: for I am sick of love," and in 7:8: "I said, I will go up to the palm tree, I will take hold of the boughs thereof: now also thy breasts shall be as clusters of the vine, and the smell of thy nose like apples."

Still further, in 8:5: "Who is this that cometh up from the wilderness, leaning upon her beloved? I raised thee up under the apple tree: there thy mother brought thee forth: there she brought thee forth that bare thee."

Judging from such pleasant passages, it would appear that apples have long been loved by the people who lived in the land of the Bible. There are, however, several factors that emphatically indicate that apples were not the fruit of the scriptures.

Botanists have determined that the native roots of apple trees, *Malus pumila*, are in the Caucasus Mountains. Although this area is within a caravan's reach of the Holy Land, it has been established that there are no common apple trees native to Israel. Apples are, in fact, a comparatively recent introduction to that part of the world.

Although apples can be grown in some parts of the Holy Land, in its broadest meaning, apple trees do not enjoy that type of climate. In fact, traveling across the United States, you will realize that apple country is in reality the northern, cooler areas. From Virginia to New England, and across the Great Lakes to our productive apple belts of Colorado and the Pacific Northwest, apple trees thrive. It would seem logical that if these trees do not prosper in hot, arid climates, the apple could not have been a tree found in the Holy Land.

Some authorities point to quinces, since they can tolerate more hot weather than apples. Others say that "apples of gold" are really oranges, and use as their proof the extensive groves of this citrus fruit that cover acres of modern Israel. In Proverbs 25:11 you will find that reference: "A word fitly spoken is like apples of gold in pictures of silver."

Botanically speaking, oranges are not indigenous to Israel. This fruit is a native of China, and the bitter Seville orange, which also has been claimed to be the "apple," is native to India. Other so-called experts identify the "apples" of the scriptures as the citron, *Citrus medica*, with its rich color and fragrant aroma. Again, the citron does not fit within scriptural context, since the trees are small and slender, most unsuitable for "sitting under" as the scriptures suggest.

Considering all the contradictions raised, especially the botanical factors and cultural requirements, there is only one fruit which appears to fit the description of the "apple." That fruit is the apricot, *Prunus armeniaca*.

Not only is the apricot found abundantly in the Holy Land, it prospers in lowlands and highlands alike, from the banks of the Jordan to the heights of Lebanon, both wild and under cultivation.

With pale leaves, this could indeed be the "apples of gold in pictures of silver." The fruit of apricot trees is strongly flavorful, sweet and aromatic as well. In Greece, and especially in Cyprus, apricots are called golden apples; again, a close proximity to that scriptural passage.

Although we realize that debate will probably linger on about the true identity of the "apple" of the Bible, the preponderance of evidence, scholarly and botanical alike, point toward the apricot as being the "apple" of the scriptures.

Apricots, the fruit of *Prunus armeniaca*, are, like peaches and plums, stone fruits. In tree growth, flower and fruit characteristics, apricots are midway between these two fruits. The trees are fairly large and spreading. In this characteristic they are much like peach trees. The leaves are broad, heart-shaped, and grow erect on the twigs. Apricot flowers are white when in full bloom, and the trees are self-fruitful. Blossoms pollinate each other without need for cross-pollination from another tree. The fruit, as all who have tasted it will probably agree, is luscious, smooth, juicy, and sweet to eat.

These trees are now cultivated widely in all of central and southeastern Asia, southern Europe, and the southern areas of the United States.

Since there still is a difference of opinion on the true identity of the Biblical "apple" trees, this chapter will provide you with a tasty choice. First, you can learn to grow apricots as multipurpose parts of your home landscape. You'll also find cultural advice for raising apples too, if you feel that you wish to accept the old English translation of the scriptures to mean apple as we know it in the United States.

Apricots prefer well-drained, light- to medium-textured soils of reasonable fertility. They bloom early in the United States, so pick a site with good air drainage to avoid spring frost damage. Most apricots are self-fruitful, but it is wise to plant two varieties, especially with the handy, dwarf-size trees, to ensure maximum fruit set. Since apricots are more tender than peaches, it is best to grow them only in moderate climates, unless you can protect them from cold and prevailing winds. New dwarf varieties have been introduced that enable you to grow full-size apricots on small trees in tubs and barrels, a most fruitful experience on your balcony or rooftop in the city.

You have a reasonable choice of modern varieties, especially from Stark Brothers nurseries, the oldest fruit specialist nursery in America. Wilson Delicious yields large fruit with luscious orange-yellow flesh. This tree is very hardy and bears large crops where others may fail. Hungarian Rose apricot closely resembles the famous Wilson Delicious. It is richly flavored and fine for eating fresh, freezing, or canning. Earli-Orange is the best early variety and Stark Giant Tilton apricot is so outstanding that this newly introduced variety has a U.S. plant patent.

Chinese Golden, available as a standard tree or a dwarf version, is very hardy, producing large fruits with firm, juicy flesh and delectable flavor. Moorpark, also in standard or dwarf size, has extra large fruits and is juicily delicious for fresh eating, cooking, or making jam.

One of the most important steps in fruit growing is proper planting. Ideal location is in well-drained soil where the trees will receive full sun daily. Dig the hole for your intended tree wider and deeper than the root ball. If you order trees by mail, they often arrive bare-rooted. Be sure the hole you provide allows sufficient room for the roots to spread naturally, as they did in the nursery.

To help trees root fast and well, improve the soil from the hole you dug if it is too heavy or too sandy. Mix in generous amounts of peat moss and humus with that garden soil.

When your trees arrive, place a mound of soil in the bottom of the hole if you are planting bare-rooted stock. It also pays to keep roots moist with wet burlap or in a pail of water to prepare the roots for their starting right in their new location.

If the trees you bought are container-grown or balled and burlapped, gently lower the root ball into the hole after you remove the container or cut the burlap wrapping free. Then fill the hole only half full of soil. Tamp it down thoroughly and add a bucket of water. When the water has drained away, fill the remaining space with soil, tamp it again to eliminate air spaces, and water well again. Leave a shallow saucer-shaped depression around the tree. That will direct rainwater and the sprinklings you provide toward the roots.

Apricot roots can be quite tender, so don't mix manure or fertilizer in the soil with which you refill the hole. Hold off adding any fertilizer until the trees have been established a month or so. Then you can spread a cupful of 16-6-4 in a ring around the trunk and water it into the soil.

If you are planting semi-dwarf or dwarf trees, understand that these are grafted trees. The desired variety has been budded to a hardy, sturdy rootstock which imparts the dwarfing effect to the size of your tree. Be sure that this graft union is 1 to 3 inches below the soil. You would plant differently for other types of fruit trees, such as apples and pears. For those, keep the graft point above the ground. Apricots prefer this in-ground planting method. If suckers rise from below the bud union, simply prune them away. For dwarf apricot trees to yield their best fruits in tubs or buckets, plant them in a fertile mixture of equal parts peat moss, loamy garden soil, and humus.

As your apricots begin to grow, your local garden center can provide you with the proper balanced fertilizer to insure they get the nourishment they need to thrive and reward you with abundant yields. Best rule of thumb is to follow the directions on the particular bag of fertilizer that you buy. The amount to use is based on the age and diameter of the tree trunk. Remember that a little plant food goes a long way, so feed your trees sparingly; only what they need.

Apricots, believed by most authorities and experts to be the "apple" mentioned in the scriptures, can reward you in wonderful ways. Not only do they have their roots in Biblical lore, but they provide sparkling white blooms in spring, with tasty eating to follow in the summer.

Apples are appealing, whether you think of them as the fruit mentioned in the Bible or merely wish to enjoy them in your fruitful landscape. Fortunately, the stigma attached to the apple which Eve picked and ate and gave to Adam to eat too has not influenced this fine fruit's popularity. Apples are as American as apple pie, it seems. In fact, the legends about this tasty fruit trace their roots to the early settlers, and to the years on the trail of good old Johnny Appleseed.

Not too many years ago, many homes had apple trees in their yards.

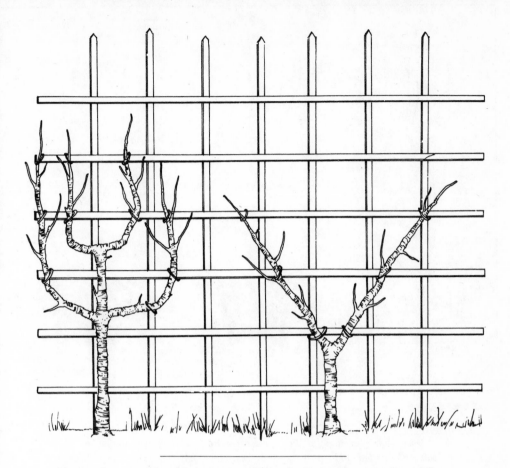

If you prefer to grow dwarf fruit trees as a property border, you can train them against a fence like this. With careful pruning, you can even create dramatic espalier effects which are attractive and distinctive. (Drawing by D. R. Sabako)

Today, largely because of urban sprawl, old productive orchards have been cut down and housing developments have sprouted in their place. However, you can begin reversing that trend with apple trees in your home landscape. Today, new semi-dwarf and dwarf-size trees have been perfected. They trace their ancestry far back to the stock derived from apples of the Caucacus Mountains. But today's apples are far superior in size, taste, and appeal. More important to your plans, dwarf-size trees do yield full-size fruit, and in greater abundance than you might imagine, while requiring far less growing room.

Luckily, many of the old favorite varieties that thrive productively in the United States have been preserved by farsighted nurseries. You have a tasty selection from which to pick your apple pleasures. In my book on

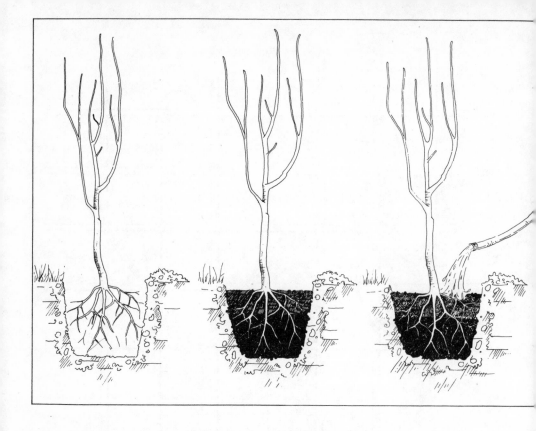

Start your young trees off right. Dig the hole large enough so roots spread naturally. Fill the hole with enriched soil mixture containing compost, humus, peat moss. Tamp soil down well to eliminate air pockets, then thoroughly soak the newly planted tree root area, and water weekly until roots begin to grow well. (Drawing by D. R. Sabako)

fruits, nuts, and berries, *Landscapes You Can Eat*, you'll find dozens of recommended varieties. Fruit nurseries like Stark's, Gurney, Henry Field, Bountiful Ridge, and others have a long list of suitable varieties.

The variety you select will depend, of course, on whether you prefer apples for eating fresh or wish to use them for pies and cooking. Here are some of the better varieties most recommended for their productivity and hardiness.

Starkspur McIntosh is renowned for its traditional McIntosh flavor and its hardiness as a highly productive tree. This variety is compact and bears heavy crops of fruit with an aristocratic aroma and tangy taste from the brilliant red apples. Starkspur Red Rome Beauty is an improved bright-red variety designed for home orchards. It is one of the best for baking, too. Ozark Gold is a yellow apple, similar to Golden Delicious, with crisp flesh and creamy rich flavor. Tropical Beauty bears well in southern climates that don't have the dormant period which most apple trees need for best performance. Red Bouquet Delicious is ideal for general home use. Its dwarf-size trees produce exquisite blooms of azalea-red, instead of white blossoms, and the fruit is flashing red, large and spicily sweet. There are many other popular varieties from Northern Spy to Lodi which also are now available on the smaller, more conveniently sized trees for home gardens.

When you're ready to plant, dig the hole for your new trees larger than you think you'll need. That way you'll be certain that roots can spread properly. If soil is poor, improve it by adding peat moss and humus to loosen clay-type soils and increase moisture-retention capacity of sandy soils. For apple trees, give full dwarf types 10 to 12 feet of growing space, semi-dwarf 20 feet, and full-size standard trees 35 to 40 feet from each other, nearby buildings, or competing trees.

When you are ready to plant apple trees, follow the directions that apply to apricots, covered earlier in this section. For tall saplings, bracing with stakes or rubber-hose-covered guy wires will let them get a firm roothold, despite winds that try to topple the newly planted trees.

Apple trees can be remarkably productive, but you must feed them so they in turn will feed you. Commercial fertilizers can be applied by the cupful around newly planted trees and scratched into the soil. For older trees, use tree-feeding spikes, those handy, nutrient-filled spikes you merely pound into the ground in a circle around your tree. Root feeding is easy too. Merely insert the required number of nutrient pellets into the hose-end device, insert the hollow, pointed tube into the ground, and turn on the water from the connecting hose. The water flow dissolves the fertilizer and carries it into the root-feeding zone underground.

Whether you elect to grow the apricot or apple as your choice for one of the fruit trees of the Bible, pay attention to its pest control. Insects and diseases can infest your trees, especially in warm summer weather. Multipurpose pesticides, however, are readily available and easy to apply. Be certain you follow all the cautions on the package label in order to win safely the battle against the bugs.

F I G S

"And the eyes of them both were opened, and they knew that they were naked; and they sewed fig leaves together, and made themselves aprons." Genesis 3:7

"A land of wheat, and barley, and vines, and fig trees, and pomegranates; a land of oil olive, and honey . . ." Deuteronomy 8:8

"And Judah and Israel dwelt safely, every man under his vine and under his fig tree . . ." I Kings 4:25

Perhaps the fig tree never enjoyed the same high esteem as the olive in the life of the people of the Bible, but it nevertheless figures prominently in the scriptures. Fig fruit played an important part in the diets of those early settlers in the Holy Land. Figs were eaten fresh during the two months that the trees produce their fruit, as they are today. Later, figs were dried to provide nourishment and needed sweetness for the remainder of the year when the land did not produce so abundantly.

You will note in the book of Judges that the fig tree does lay claim to high esteem since it is mentioned along with the vine and the olive as a productive tree indeed. In Proverbs 27:18 it is written: "Whoso keepeth the fig tree shall eat the fruit thereof . . ." and in Isaiah 36:16 the fig is valued as a desired food: "Hearken not to Hezekiah: for thus saith the king of Assyria, Make an agreement with me by a present, and come out to me: and eat ye every one of his vine, and every one of his fig tree, and drink ye every one the waters of his own cistern."

Again in Joel 2:21–22, the fig is mentioned in the same breath with the productive, highly valued vine for its grapes and wine: "Fear not, O land; be glad and rejoice: for the Lord will do great things. Be not afraid, ye beasts of the field: for the pastures of the wilderness do spring, for the tree beareth her fruit, the fig tree and the vine do yield their strength."

The fig tree of the Holy Land produces its crops in a short two-month season—in some cases two crops a year. In the parable from St. Matthew, the devout were given another message about the seasons in 24:32: "Now learn a parable of the fig tree; When his branch is yet tender, and putteth forth leaves, ye know that summer is nigh." This message was said another way by St. Luke in 21:29–30: "Behold the fig tree, and all the trees; When they now shoot forth, ye see and know of your own selves that summer is now nigh at hand." Palms and olive trees may be more frequently associated in our minds with pictures of the land of the Bible as we read the scriptures, but the fig, *Ficus carica*, is without doubt equally significant. It is mentioned more than fifty times and also is the first plant to be mentioned by name in the story of the aprons made from fig leaves in the Garden of Eden. From Genesis through the many books of the Old and New Testaments, these marvelous trees and their rich fruits obviously had won the admiration and appreciation of the children of Israel and those early Christians who followed.

In Nahum 3:12 the sweetness of the tempting figs is again recorded: "All thy strong holds shall be like fig trees with the firstripe figs: if they be shaken, they shall even fall into the mouth of the eater."

References to figs are too numerous to list completely here, but with the help of the Concordance you can pursue the many mentions of them in the scriptures. They may not be the most imposing trees, nor perhaps as valued in commerce in olden times as the olives, but they were undoubtedly one of the more reliable plants of the Bible that gave forth their fruit in season.

The fig is considered a native of southwestern Asia and Syria. However, in early times as today it was cultivated extensively in Egypt, Palestine, and the surrounding areas. If you were to visit Israel today, you would find both wild fig trees and hundreds after hundreds of acres of cultivated fig orchards. In fact, most private homes have a fig tree growing in the garden. This wide popularity of the fig tree today can be compared with the Biblical passages from Zechariah 3:10: "In that day, saith the Lord of hosts, shall ye call every man his neighbour under the vine and under the fig tree," and in Micah 4:4, where it is told: "But they shall sit every man under his vine and under his fig tree; and none shall make them afraid . . ."

As you read the scriptures, you may be puzzled by references to sycomore trees. In I Chronicles 27:28, you will find this passage: "And over the olive trees and the sycomore trees that were in the low plains . . ." and later, in II Chronicles 1:15, you will find references to "cedar trees made he as the sycomore trees that are in the vale for abundance." Still later, in 9:27, there is this passage: "And the king made silver in Jerusalem as stones, and cedar trees made he as the sycomore trees that are in the low plains in abundance."

Familiar as we are with the sycamore tree in Europe and the United States, it is easy to be misled by these references. But the sycamore tree with which we are acquainted is in reality the plane tree, *Platanus occidentalis*, or London plane tree, *P. orientalis*. These have no relation to the sycomore tree of the Bible.

That tree of the scriptures is really the sycamore fig tree, *Ficus sycomorus*. This widely spreading tree may rise only 40 feet from the earth, but its sturdy trunk can support sweeping branches 80 to 100 feet in diameter. Although common in Egypt and Israel, this tree differs from the common fig in several ways. It is usually evergreen and has much smaller, unlobed leaves. Growing abundantly in the lower valleys, it does indeed provide delightful shade along roadsides in the hot Mediterranean climate, and it may be the tree mentioned as suitable for sitting under.

Too large for convenient growing, it is mentioned here to put these passages about the sycamore tree into their proper perspective.

Fig trees vary in their growth habit. Some may seem no more than straggly shrubs sprawling from their roothold in stony areas or rocky walls. In more favored growing situations, fig trees may grow 20 to 30 feet tall. These trees have broad and somewhat rough leaves, deeply lobed or oval-shaped.

Through the ages, figs were prized food among the Greeks and

Romans too. Even in Roman legends, the fig tree that stood over the twin founders of Rome in the wolf's cave became an emblem of the prosperity of the Roman Empire.

Today, figs are still widely grown in many countries surrounding the Mediterranean Sea. They also are grown commercially in the United States, but primarily in the southern areas, Texas and California. Most find their way to the canning industry, rather than the dried products which were so popular among peoples of the Old World. In Mediterranean countries, figs are so widely used, both fresh and dried, that they have been termed the poor man's food.

Figs have an unusual habit of producing fruit buds before the leaves have emerged from the dormant period. Perhaps this is one explanation of the story in St. Matthew concerning the coming of the summer season. When new leaves are fully out, the fig fruit should normally be ripe. In this respect, figs produce their fruit in reverse of the pattern of most fruit trees.

As you read the scriptures, you will note that when the prophets berated their followers, they frequently threatened that the vine and fig crops might be destroyed because of their people's wickedness. In other passages, especially the peaceful verses recalling how pleasant it would be to sit under one's own fig tree, you can realize that the fig tree had many meanings, including its symbolism of peace and plenty.

In most parts of the United States, fig trees will not prosper because they cannot tolerate cold winter weather. If you are fortunate to live in the most southern states, however, you can enjoy true fig trees, perhaps in your dooryard. Fortunately for northern Americans, there are related ficus trees that can be grown easily outdoors or inside, in homes, offices, and even modern shopping malls. Those you can enjoy as representative of this important plant of the Bible.

Among fig trees, you actually have a delightful assortment of decorative trees that trace their ancestry to the land of the Bible.

Actually the *Moraceae*, known as the fig or mulberry family, includes about a thousand species, from woody trees to tiny vines. Best known in the family are the fig and the breadfruit.

In the Bible, fig trees are often mentioned, since the figs that grew in the Holy Land were as vital to the survival of the people there as dates from the palm trees. You may not have an opportunity to grow the large fig tree, *carica*, with its abundant crops of edible figs, but you can still feature fig trees in your life.

For ornamental use to represent the Biblical fig trees, you can try *Ficus benjamina*, a weeping fig tree widely available for indoor decor. *Ficus lyrata*, the fiddle-leaf fig, also will thrive in your home in pots and tubs. *Ficus carica*, the common fig, and *Ficus cyathistipulata* both can bear fruit if

Ficus benjamina, *a decorative relative of the fig trees of the Bible, performs well as an attractive house or office plant. You may prefer single- or multiple-trunk type. (Photo courtesy of Florida Foliage Growers)*

they are provided with the proper combination of sun, moisture, and nourishment. Remember their original habitat. Fig trees and other members of the ficus family need lots of sun, at least 6 to 8 hours daily, to hold their foliage and prosper.

You will find a variety of ficus trees available through florists and leading garden centers as well as through mail order nurseries. In fact, the Florida Foliage Growers have recently embarked on a major educational program encompassing the fig tree and other plants of the Holy Land. Whichever fig tree you decide to grow as your own symbol of the fig tree of the scriptures, here are basic cultural needs which will help them thrive.

Fig trees are evergreen with glossy green, somewhat leathery foliage, which varies by species. Most are easy to maintain in containers and are surprisingly durable. If they grow too large for their location, you can easily prune them back to a more desired shape and size. Because they are so versatile, ficus trees have been well used in shopping malls, offices, and other indoor plantscapes.

Fig trees do best in a relatively small container filled with soil mixed with peat moss. This allows their roots to drain off excess water. All roots must breathe, as you know, especially when restricted to the environment of a container. Ficus trees enjoy moisture. It is usually best to let the soil become slightly on the dry side; then soak it thoroughly. If overly wet, roots rot and black spots appear on leaves. During winter, when sunlight is low, some fig trees drop leaves. If you are providing proper moisture and the plant container has satisfactory drainage, a little shedding of leaves may not signal any problems. Fig trees do have a tendency to drop some leaves at times.

If you are fortunate to live in far southern areas of the United States, rejoice. You can most likely grow an everbearing fig, representative of the fig tree of the scriptures. The Texas ornamental fig tree grows as a shrub or small tree in horticultural zones 8 through 10. This specimen bears prolifically in the late summer to late fall, fruiting around September in zones 6 and 7. This ficus will need protection from freezing, of course, especially where temperatures fall below 10 degrees F. It is best planted in tubs so that the plants can be moved to a cool, moist, indoor environment during late fall and winter. The Texas everbearing fig tree is self-pollinating.

Another related fig, called Brown Turkey, is an everbearing type that is ideally suited for growing in containers or tubs. It will winter safely in a cool cellar, or you may allow it to harden off with a few light fall frosts. Then, you can bring it indoors for growing as a houseplant during the winter. If, accidentally, the top is frozen, you can simply cut the tree to the ground. When it begins growing again with the arrival of properly warm weather, it will produce fruit on this new wood. Brown Turkey variety of fig also can be grown in the open ground, and will develop a crop of tasty fruits as far north as Maryland. Plants usually bear the first year, producing fruit that is very sweet and of good size. The flesh is firm, meaty, and delicious. This variety is suggested by Wayside Gardens for culture in zones 6 through 10. Both these figs prefer moderately fertile soil, good sun, and ample water, especially during the fruiting period.

GRAPES

"And Noah began to be an husbandman, and he planted a vineyard: and he drank of the wine . . ." Genesis 9:20, 21

"And the chief butler told his dream to Joseph, and said to him, In my dream, behold, a vine was before me; And in the vine were three branches: and it was as though it budded, and her blossoms shot forth; and the clusters thereof brought forth ripe grapes . . ." Genesis 40:9–10

"And sow the fields, and plant vineyards, which may yield fruits of increase." Psalms 107:37

"The fig tree putteth forth her green figs, and the vines with the tender grapes give a good smell . . ." The Song of Solomon 2:13

Grapes, vineyards, and their precious products account for more passages about plants in the Bible than any other plant. That is not surprising, considering the importance of this fruit to the peoples of the Bible. From the days of Noah, when he became a husbandman and began to till and tend the land, grapes have been a life-sustaining crop. Not only do they furnish food in the form of fresh and dried grapes, but their juice has been favored through the ages. Wine is still preferred over water as a daily beverage in some countries, such as France, Spain, and Italy.

The common grapevine, *Vitis vinifera*, has sent its roots into the soil of many lands since the days of the Bible. These plants have been cultivated by the human race through the courses of so many civilizations that the exact native region is unknown. Seeds found in Swiss lake dwellings of the Bronze period and others discovered in the tombs of Egypt closely resemble seeds of the grapes grown today.

Some botanists believe that the grape actually originated in the hilly regions of Armenia and other countries bordering the Caspian Sea. No precise determination has been made and probably never will be. In early Egypt, in the Fourth Dynasty and in the later Seventeenth and Eigh-

Grapes grow well in most parts of the United States, producing their luscious clusters for fresh eating or for making juice and wine. (Photo courtesy of R. T. French Company)

teenth dynasties, details about grapes and winemaking were inscribed in mosaics and on the walls of Egyptian tombs. The Fourth Dynasty, of course, places grape culture knowledge at about 2400 B.C.

Perhaps because grapes were first mentioned in the story of Noah in Genesis, he is credited with the introduction of grapes into cultivation. We know from ancient records that they were well known and valued in Assyria and Egypt. Inscriptions and sculptures salute the grape and its wine as far back as written records exist. Archaeologists also have uncovered the remains of winepresses in the rocks of Palestine, attesting to the fact that these people of the Bible days gathered their grapes and crushed them for the juice thousands of years ago.

Through the ages since, grapes have been a part of every civilization. Today, grape cultivation continues throughout the world, especially in those regions where the soils and climates favor this productive vine's growth. Wherever grapes originated, we are able to trace them to the

Holy Land in Psalms 80:8–10: "Thou hast brought a vine out of Egypt: thou has cast out the heathen, and planted it. Thou preparedst room before it, and didst cause it to take deep root, and it filled the land. The hills were covered with the shadow of it, and the boughs thereof were like the goodly cedars."

These passages can be understood in several ways. They can explain how the cultivated grapevine was brought by the children of Israel out of Egypt into Palestine to replace the wild plants that already grew there. You can also consider this in terms of the spreading and transplanting of the tribes of Israel into their Promised Land. Biblical scholars point out that the terms "fruitful vine" and the "vine brought out of Egypt" are symbolic of the Jewish people, moving from their period of bondage into the land of the Bible.

Grapevines are the first cultivated plant mentioned in the Bible, as the passages about Noah's becoming a husbandman suggest. From that passage in Genesis, through Exodus, Leviticus, Numbers, and most of the books of the Old Testament, you can readily find other mentions lauding the grape or its juice, or using this most widely cultivated plant of the Bible symbolically.

Wine, of course, was a traditional drink throughout the land of the Bible. Not only was it used as a beverage, but wine also was employed in ceremonies, along with oils and spices in the temples of the people. When the Roman Legions were in the Holy Land, they too savored the grapes and the wine, no doubt carrying back cuttings to plant the vineyards that grew throughout ancient Rome. Before them, the Phoenicians had taken the grape to France, probably about 600 B.C. The Greeks enjoyed the grape and its vintage products, and no doubt also helped expand its cultivation during their golden age.

The Romans are credited with bringing the grape to England during their conquests there. At the same time, caravans of commerce extended the range of the grape into the Orient, where it flourishes today. Among all of mankind's cultivated fruits, the grape is probably the most widely grown in the nations of the world today.

Grapes were often mentioned symbolically according to the scriptures. Sharing one's abundance is urged as a duty of the faithful, as you can read in this passage from Deuteronomy 24:21: "When thou gatherest the grapes of thy vineyard, thou shalt not glean it afterward: it shall be for the stranger, for the fatherless, and for the widow."

The importance of grapes to the people of the Bible also can be seen in other passages. In Deuteronomy 20:6, Moses exhorts his people to prepare for battle, but also mentions the need to tend the land and crops: "And what man is he that hath planted a vineyard, and hath not yet eaten of it? let him also go and return unto his house, lest he die in the battle, and another man eat of it."

As you read on in the New Testament too, grapes, vineyards, and wine are almost as prominently mentioned as they were in the books of the Old Testament. Not only were grapes and the vineyards employed symbolically in parables, but the importance of wine was given its most significant place in the Christian heritage and tradition.

In the story of the Lord's Supper, St. Matthew marks the importance of wine to Christians for all time in 26:26–29: "And as they were eating, Jesus took bread, and blessed it, and brake it, and gave it to the disciples, and said, Take, eat; this is my body. And he took the cup, and gave thanks, and gave it to them, saying, Drink ye all of it; For this is my blood of the new testament, which is shed for many for the remission of sins. But I say unto you, I will not drink henceforth of this fruit of the vine, until that day when I drink it new with you in my Father's kingdom."

With these thoughts in mind as you reread the scriptures, perhaps grapes should be first on your list of plants of the Bible for planting in your home grounds. This chapter will tell you how to grow them successfully.

The common grape of the Old World is *Vitis vinifera*. Its leaves differ from most of our cultivated American grapes. They are rounded, heart-shaped, and 5-lobed, with coarse teeth. The branches bear tendrils and small, inconspicuous greenish flowers. In ancient times, these grapevines often attained great height from their gnarled old trunks. Travelers in Israel today often remark at the size of some of these old vines, climbing to the second story of a home or building.

Old World grapes were brought to the United States with the first settlers, and for a time they prospered. Unfortunately, a highly destructive insect in this country also found the grapevines good to eat. The grape phylloxera, a tiny plant louse, virtually destroyed the grapes that were brought by the colonists and established originally in America.

At one time, this pest was blamed for destroying most of the old rootstock grapes that existed. Fortunately, native species of grapes had developed a resistance to this pest and diseases that infected the vines. Using these native species, notably *Vitis labrusca* and variations of it, plant breeders created hardier, resistant plants. These developments revitalized the grape industry and made it possible for grapes to be grown across our continent.

Plant pests have a way, as locust and caterpillar plagues of Biblical times did, of traveling vast distances to attack their favorite host crops. Phylloxera was introduced to Europe on stocks of American grapevines exported there, and this pest soon threatened much of Europe's grapevines. Eventually, using the resistant rootstocks developed by American viticulturists, the industry was saved. Today it flourishes again, with many vineyards thriving on grapevines from the rootstocks developed in

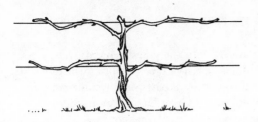

FOUR ARM, SINGLE TRUNK KNIFFIN SYSTEM.

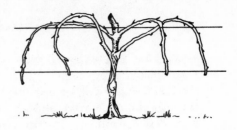

FOUR ARM, SINGLE TRUNK UMBRELLA SYSTEM.

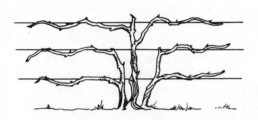

SIX ARM, THREE TRUNK, THREE WIRE,
MODIFIED KNIFFIN SYSTEM.

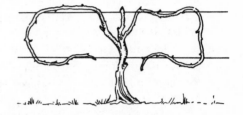

TWO ARM, SINGLE TRUNK UMBRELLA SYSTEM.

If you wish more dramatic displays from grapevines, train them on a fence as shown in these illustrations. Remember that you must prune your vines every year to encourage them to produce most abundantly for you. (Drawing by D. R. Sabako)

America. Today, a variety of improved rootstocks are utilized both in this country and abroad.

As you plan to grow grapes representative of their place in the scriptures, perhaps some brief background on the various types will be helpful. Grapes are grouped according to their use as wine grapes, table grapes, sweet juice grapes, and those for making raisins or canning. There are some eight thousand different varieties known today. Only a limited number, however, are suitable for high-quality wines, although any grape can be crushed and its juice fermented. Many are, of course, ideal for growing in your own home grounds for several purposes.

Grapes, being vines, offer unique qualities for your garden plantscaping. They'll grow up trellises, fences, and poles to save space in your garden ground. You can also train them into unique and dramatic designs

if you wish. You may prefer a grape arbor, with the vines twining over-
head, where you can sit in the shade, plucking tasty grapes at your leisure.
Nurseries offer a wide choice of varieties that fit all parts of our country.
Whether you live in the Gulf Coast states or the Great Lakes region, there
are varieties suitable for your garden.

Grapes prefer abundant sun, well-drained soil, and good air circula-
tion. Air movement is more important with grapes than other fruit crops
because it helps prevent certain diseases that may affect them. Fortunately, new varieties have built-in disease resistance, and modern fungicides enable you to combat any problems that occur.

You can choose red, white, pink, or purple grapes, for fresh use, jam,
jelly, and wine as well. Mail order catalogs list many, but some of the best
are included here. Grapes are, as they have proved for centuries, sturdy,
stubborn, and amazingly long-lived plants. Some vineyards have been
producing prolifically for decades, even centuries in Europe. As you pick
the right spot for your vines, remember they will be serving you for years
to come, so plant and tend them properly.

Grapes can withstand drought and cold and can succeed even in
rather sandy, seemingly poor soil. With extra care to improve their growing conditions, they will reward you more generously. You can improve
soil by mixing compost and well-rotted manure with it. To prepare a
planting site, place these materials on the ground, then spade or rototill
them under, incorporating this helpful organic matter into the existing
soil. Doing this in the fall prior to planting improves the planting site
measurably.

When you order plants, buy vigorous two-year-olds. They cost a bit
more but will give you a better start on your grape-growing activities.
Grapevines, you should know, are somewhat slow starting, so sturdier
vines initially are an asset.

If you live in southern areas, try these fine bunch-type grapes.
Among blue-black types, Buffalo has excellent quality for table use or
winemaking. Van Buren has hardy, vigorous vines that produce tasty
table grapes. Steuben and Concord ripen in midseason and are the most
popular varieties.

Among White types, try Himrod, Interlaken, or Seneca. All are
early and have high dessert quality. The first two are seedless. Among
Red varieties, Delaware has high sugar content but vines lack vigor.
Catawba bears late, but its vigorous vines yield large, tasty berries for
fresh use or wines.

For northern areas with climates similar to the Great Lakes region,
here are recommended varieties. Niagara is the most popular for its large,
compact bunches of fruit. Red Delaware is fine for table and making wine.

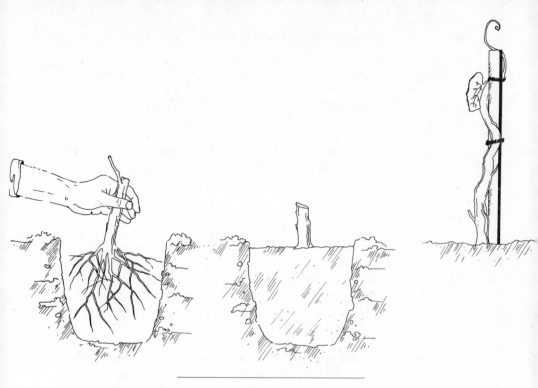

1. Place the grapevine in a hole large enough to let the roots spread naturally.
2. Fill the hole with improved soil mix, tamp it down well to eliminate air pockets, and water thoroughly so the vine gets a proper, healthy start.
3. As your vine begins to grow, train it to a post, fence, or trellis for support. (Drawings by D. R. Sabako)

Blue Concord and Red Catawba also are good choices. Buffalo, Agawan, and Stark Blue Boy also perform well where hardiness is a factor.

New grapes, bred especially for winemaking, include Cascade, a hardy and productive sweet grape, Baco Noir, which produces excellent burgundy wine, Foch, which becomes a zesty red wine, and Seyval Blanc, which produces well and yields a white table wine.

To plant your young vines, open a hole in the ground large enough to accommodate root spread naturally, 15 to 18 inches across. Set the vines the same depth as they grew in the nursery, which you can judge from the markings on the vine itself. Be sure to keep the graft point of grafted grapes above the ground to discourage unwanted sucker growth. Add soil and firm it around the roots. Add a half bucket of water and fill the remaining space with soil. Water again and leave a saucer-shaped depression around the plant. This will direct rain toward the newly forming feeder roots.

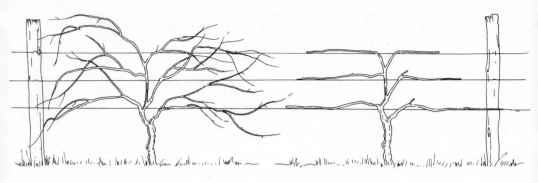

A grapevine that is overgrown is not productive. Prune away extra canes to force new fruiting wood growth. Your vines in spring should be cut back as shown on the right.
(Drawing by D. R. Sabako)

Although grapes will grow under dry conditions, watering weekly, especially as the fruit is forming, encourages greater yield of larger, juicier grapes. The first year after planting you can fertilize your vines with a quarter pound of 10-10-10 spread in a circle around each vine. Repeat this application monthly until midsummer. Mulching with compost or rotted manure is beneficial too. In future years, you can increase amounts. For bearing vines, after the first two seasons, spread 2 to 4 pounds of well-balanced plant food around each vine. You'll find more details about application rates for various plants on the fertilizer containers themselves, of course.

Grapevines have both shallow roots and deeper feeding ones. If you cultivate around your plants, don't dig deeply. Grapevines will let you know when they are well nourished. Vigorous growth and nicely plump buds indicate the plants are satisfied. Your fruiting canes should be ¼ to ½ inch in diameter if the plants are properly fed.

Grapes aren't too fussy about the type of support they have, but they do best when tied to a trellis, arbor, or wire fence that provides good air flow around them. You can build a sitting arbor of cedar or redwood posts with a lattice work top. As your vines grow tall, train them up the posts by tying them for support. Then, the fruiting vines can wander over the top, providing you with shade as a bonus in your backyard sun.

Heavy-duty wire strung between wood or metal posts also works well for property border plantings. Since grapes live a long time, it pays to erect sturdy, permanent supports for them. The illustrations here show you how to prune your grapevines on a fence. The trunk is the permanent part of your plant. The arms of the vine are short side extensions of this main stem. The shoots on your vines arise from buds on the fruiting wood and bear the leaves, flowers, and eventually the grapes themselves. With grapes, fruiting shoots and canes are the one-year-old growth that yields the fruit.

Pruning is important. It actually strengthens the vine, encouraging it to form new fruiting canes for more abundant harvests.

Your objective should be to train the vines so that their arms and fruiting canes follow the wires of your fence or trellis. Strange as it may seem, you only need two or three canes at each wire to produce the fruiting wood that bears the grapes. Since grapes sprout new shoots each year, most grapes are borne on the fruiting canes the second season. By removing old tired canes regularly, you encourage the greatest productivity from your vines.

Whether you wish to read every scriptural passage that refers to grapes, the vines, and wine or not, one thing is certain: sitting under your grape arbor plucking the tasty fruit from the vine can make your life more pleasant, as it was for those who did the same in Biblical times.

OLIVES

"And the dove came in to him in the evening; and, lo, in her mouth was an olive leaf pluckt off: so Noah knew that the waters were abated from off the earth." Genesis 8:11

"Thou shalt have olive trees throughout all thy coasts, but thou shalt not anoint thyself with the oil; for thine olive shall cast his fruit." Deuteronomy 28:40

"Thy wife shall be as a fruitful vine by the sides of thine house: thy children like olive plants round about thy table." Psalms 128:3

"But I am like a green olive tree in the house of God: I trust in the mercy of God for ever and ever." Psalms 52:8

In the scriptures, the olive tree vies with the grapevine for the profusion of references made to it throughout the Bible. There is no doubt that the olive tree was one of the most valuable trees to grace the Holy Land. From it, the people obtained the fruit and its oil, both of great importance to their daily lives, their worship, and the trade that the early people of the Bible had with other peoples. The wood of the tree itself was used for furniture as craftsmen took advantage of its fine grain and rich, variegated color.

Wood of olive seems to have had strong appeal to craftsmen of Biblical days, as indicated in this passage from I Kings 6:31–33: "And for the entering of the oracle he made doors of olive tree: the lintel and side posts were a fifth part of the wall. The two doors also were of olive tree; and he carved upon them carvings of cherubims and palm trees and open flowers, and overlaid them with gold, and spread gold upon the cherubims, and upon the palm trees. So also made he for the door of the temple posts of olive tree, a fourth part of the wall."

If you were to count all the references to olives and to olive trees, and add them to mentions of oil and anointing in feasts and during religious ceremonies, the total times this plant appears in the Bible boggles the mind.

In the passages from Genesis, as Noah and his family waited patiently for the rains to cease and the flood to stop, the olive is perhaps emphasized most significantly. It is from this passage, when the dove with olive branch in her mouth returns to the ark, that the symbolism of both dove and olive have emerged. Ever since, the olive branch and the dove have been associated with peace and friendship

As you read the holy words, you'll find various clues to the value of the olive tree and its products. Of course, the fruit was eaten by the people, both fresh and no doubt dried and preserved for the time when there were no crops to harvest. Wandering through the wilderness, the children of Israel longed for the foods and fruits of Egypt. When they finally were told by their scouts of the wonders of the Promised Land, its blessings seemed too marvelous to believe. Olive trees were among the fruits that grew in such abundance in the land that flowed with milk and honey.

The olive tree, *Olea europaea*, is perhaps more closely associated with the history of civilization than any other tree. It has nourished multitudes for countless centuries and been the basis for much of the trade that began early in mankind's history. Today the olive tree and its precious fruits remain a major crop for a number of countries, especially Spain, the world's leading olive-producing nation.

Search you may, but the origin of the edible olive, *Olea europaea*, has long been lost in ancient history. Botanists believe its native soil is in the eastern Mediterranean, or perhaps in south-central Asia. Unquestionably it is one of our world's oldest cultivated crops. Some olive orchards that still yield today, bear their olives on trees that may be more than a thousand years old.

From whatever native soil olive trees first sprang, their fame has led to their cultivation throughout the countries that border the Mediterranean Sea. This subtropical, broad-leaved evergreen tree is in the *Oleaceae* family. You may recognize some of its distant relatives in your own garden. They include the ash, lilac, forsythia, and privet.

Olive trees grow wild, of course. Probably any solitary olive trees

Olive trees were one of the most important trees of the Holy Land, providing fruit and oil as they do today in Israel and other Mediterranean countries.

that you may have seen in the Holy Land or in southern European countries are not really wild. The olive tree is one of the longest lived among trees, especially among fruit-bearing trees. They are, it is true, slow-growing, requiring many years to reach a height of 10 to 40 feet, depending on their situation. However, specimens are known that are believed to be well over a thousand years of age, based on the tracing of old records where they grow.

Despite the wars that have ravaged the land of the Bible through the ages, olive trees have persevered. Even when orchards were cut down by rampaging armies, the trees returned, sprouting up from their tenacious roots, often producing several trees where one grew before. This ability to grow back when its top is cut or dies has enabled the olive tree to attain its legendary age in many places.

The olive tree is amazing in other ways as well. It is one of the few trees that will tolerate long periods of drought, up to 5 to 6 months through summers, so long as it receives a fair share of winter rain. Small wonder that this tree was so prized and revered by the ancients. It has been suspected that because of this great tolerance for adverse conditions olive trees do not require fertile soil. That may seem true, but the fact remains that when planted in good soil and provided with ample water and nitrogen fertilizer, olive trees increase their abundance markedly.

During Biblical times, the olive tree was probably even more abundant than it is today in the Holy Land. Modern times have brought new demands for use of land. Many other crops now grow where olive orchards stood centuries ago. That fact can be proved by observing the gnarled ancient trees that stand alone near modern farming operations in Israel today.

That olives and their oil were both vitally important to the peoples of the Holy Land is obvious, as you review the scriptures. Olive oil provided fuel for lamps in the temples and in people's homes. "And thou shalt command the children of Israel, that they bring thee pure oil olive beaten for the light, to cause the lamp to burn always," it is written in Exodus 27:20. Even today in remote regions olive oil is used for lighting lamps.

In Deuteronomy, Moses tells his people of the purpose of God's laws and exhorts them to faithful obedience: "And it shall be, when the Lord thy God shall have brought thee into the land which he sware unto thy fathers, to Abraham, to Isaac, and to Jacob, to give thee great and goodly cities, which thou buildest not. And houses full of all good things, which thou filledst not, and wells digged, which thou diggedst not, vineyards and olive trees, which thou plantedst not; when thou shalt have eaten and be full." Olive trees, we must assume, were plenteous in the Promised Land, the land of the Bible, according to Deuteronomy 6:10–11.

In the teachings of the scriptures, generosity is shown to be a virtue, and sharing of fruits and crops was a worthy thing to do. You can read this admonition also in Deuteronomy 24:20: "When thou beatest thine olive tree, thou shalt not go over the boughs again: it shall be for the stranger, for the fatherless, and for the widow." These passages remind us that when we have abundance, there is a need to share with those less fortunate, for in Deuteronomy 24:22 it is well said: "And thou shalt remember that thou wast a bondman in the land of Egypt: therefore, I command thee to do this thing."

Olives and their oil were also valued articles in trade in the earliest caravans of commerce. "And Solomon gave Hiram twenty thousand measures of wheat for food to his household, and twenty measures of pure oil: thus gave Solomon to Hiram year by year." This passage, I Kings 5:11, again indicates as other verses of the Bible do that oil was one of the major articles of trade employed in the barter system used in ancient times.

Olive oil was valued also for religious purposes. Not only did it help light the temples, but it was widely used in anointing members of the faithful.

In the 23rd Psalm of David, God's tender and constant care is assured to all who obey his laws. In the fifth verse, so well known throughout the world, it is written: "Thou preparest a table before me in the presence of mine enemies: thou anointest my head with oil; my cup runneth over." Anointing with oil is mentioned in other books of the Bible too, for many occasions. In addition to the purifying value for which olive oil was used, it also was used in ointments for medicinal purposes during Biblical days and since.

Olive trees have been mentioned in the scriptures as desirable in other ways. In Hosea the return of those fallen by their iniquity was exhorted with the blessing of God promised for them who returned to the ways of the faithful. In Hosea 14:6, the olive tree is again used as a symbol of what is good: "His branches shall spread, and his beauty shall be as the olive tree, and his smell as Lebanon."

As you reflect on the passages throughout the Bible that refer to olive trees, remember that the gardens of Biblical days were not the flower gardens we may think of as we read the scriptures. They were primarily orchards, enclosed by hedges of thorn or stone walls. Usually the trees were olives with perhaps figs and nut trees scattered here and there among them or in the corners of these gardens. The Garden of Gethsemane was, in reality, an olive orchard garden at the foot of the Mount of Olives. No doubt the oil presses were located nearby, to press the precious olive oil from the fruits of these magnificently productive trees.

You can grow olive trees outdoors if you live in a climate that is appropriate for their growth, such as the Southwestern United States, southern California, southern Texas and Florida. These trees have a unique ability to survive even in poor soils and dry areas. They will, however, thrive if given the same care in improving soil, providing additional water, and applying light applications of nitrogen fertilizer around them beneath the spread of their branches.

Plant olive trees as you would other fruit trees. There are few insects that bother these trees, which is itself a blessing. You may not see the fruits of your labors in the form of olives, since these trees do need their special allotment of climatic conditions which are found primarily in their native areas. However, they are worth a try.

A few specialty nurseries may have saplings available, but since there is little demand for olive trees in the United States, it may be necessary to contact leading arboretums to locate a suitable source. Since California is the primary commercial olive-producing area in America, nurseries there are your best bet for trees.

Olive trees are started from 4- to 6-inch-long prunings, planted in sand under artificial heat. Once rooted, trees are transplanted to a nursery where they grow for two or three years and then are ready for their permanent home, your yard or a commercial orchard. Olive trees will bear in about seven or eight years from the time they are planted in the orchard or permanent location, provided the environment is suitable for them.

PALMS

"And they came to Elim, where were twelve wells of water, and threescore and ten palm trees: and they encamped there by the waters." Exodus 15:27

"And the south, and the plain of the valley of Jericho, the city of palm trees, unto Zoar." Deuteronomy 34:3

"And she dwelt under the palm tree of Deborah between Ramah and Beth-el in mount Ephraim: and the children of Israel came up to her for judgment." Judges 4:5

Palm trees have been associated with the land of the Bible from time immemorial as they are with the land of Egypt where they have provided sustenance to wanderers across the desert. There is no question in the minds of scholars that the date palm, *Phoenix dactylifera*, is indeed the palm tree of the scriptures.

Tracing the history of plants from the Holy Land, the date palm stands tall as a symbol of life in the barren wastes through which the children of Israel wandered. With other food in short supply in desert areas, it is no wonder that the peoples of that region appreciated the palm tree. It was and is a life-sustaining plant. Virtually all parts of these trees prove useful. The dates are sweet and succulent. The wood itself was used for building as its leaves were woven for many purposes, from brooms to thatched roofs of shelters. Rope is still made from the weblike portions of the palm tree's crown. Even the kernel of the date is ground up, soaked in water, and used as livestock food.

Scholars believe that at one time, before the ravages of war and drought, vast areas of the Holy Land were covered with palm trees.

The awe and reverence in which ancient peoples held this tree can be appreciated from the scriptures: "And he shall be like a tree planted by the rivers of water, that bringeth forth his fruit in his season; his leaf also shall not wither; and whatsoever he doeth shall prosper" Psalms 1:3. In Psalms 92:12–14, you can read: "The righteous shall flourish like the palm tree: he shall grow like a cedar in Lebanon. Those that be planted in the house of the Lord shall flourish in the courts of our God. They shall still bring forth fruit in old age; they shall be fat and flourishing."

The stately palms, towering over the desert, signaling to the tired traveler the presence of water at the oasis from which they obtained their moisture, evoke vivid images even today. In The Song of Solomon 7:7–8 you can read this imagery: "This thy stature is like to a palm tree, and thy breasts to clusters of grapes. I said, I will go up to the palm tree, I will take hold of the boughs thereof . . ."

Palm fronds, the leaves of the palms, were often used in important ceremonies in the days of the Bible, as you can understand from St. John 12:12–13: "On the next day much people that were come to the feast, when they heard that Jesus was coming to Jerusalem, Took branches of palm trees, and went forth to meet him, and cried, Hosanna: Blessed is the King of Israel that cometh in the name of the Lord."

In Revelation 7:9, again the symbolism of palm leaves is apparent: "After this I beheld, and, lo, a great multitude, which no man could number, of all nations, and kindreds, and people, and tongues, stood before the throne, and before the Lamb, clothed with white robes, and palms in their hands."

From those early days of the Bible, palm leaves have carried with them religious meanings. Medieval Christians, according to old texts, believed that angels carried branches of palms to smooth the way for martyrs on the road to heaven. In the stories of St. Christopher, St. Clare, and St. Francis of Assisi, palm fronds figure prominently. Their significance remains today as the palms are distributed to the devout on Palm Sunday.

With these thoughts in mind, perhaps the palm tree in one of its more convenient, smaller relatives may be among your early choices as you begin your garden of Biblical plants.

You may not realize that palm trees are evergreens. At least they are in the tropical and warm climates in which they live. The ancient palm

An Areca Palm in a container provides you with the feeling of the palms of the Bible in your indoor decorating scheme. (Photo courtesy of Florida Foliage Growers)

family actually includes between 2,500 and 4,000 species. They are ever-growing woody plants, mostly trees, but some are climbers and others bushes.

Deciding which among these palms are the true plants of the Bible may seem difficult at first. However, the date palms of the desert areas, still found along the seacoast of Israel and in oases of the deserts, leave little doubt that these were the trees of the Bible. With their nutritious fruits, they succored many peoples through the centuries before and after the time of Jesus Christ.

The date palm is *Phoenix dactylifera* and has been the tree of life of the ancient peoples from Asia Minor to Egypt for more than five thousand years. It was cultivated in Mesopotamia from Sumerian times, and its likeness is carved into the temple at Karnak in upper Egypt, which was built about 1370 B.C.

Palms are unique among trees. Actually they are monocots, close relatives of lilies and grasses, but palms have evolved into tree-type plants. The trunk of a palm is unlike our familiar trees. They have a hard central core of wood which regularly increases both in length and in diameter by its cambium layers. In contrast, most palms have only a single point of growth activity, the terminal bud. If this is cut away, the tree will die. Without a cambium layer, palm tree trunks cannot increase in diameter. Thus, they remain graceful and slender. Also, in contrast to most trees, palms have a soft spongy heart area surrounded and protected by a hard casing of strong fibers which serve as the tree's conductive system for moving water and minerals to the leaves and growing areas.

Some species, such as the date palm, carry male and female flowers on separate trees, which means cross-pollination is necessary to produce fruit. The Babylonians and Egyptians knew this and cross-pollinated their date palms centuries before the life of Christ.

The date palm, food tree of the scriptures, is too tall for indoor culture and, in reality, is not widely grown as a landscape plant outside of selected arboretums and public gardens. Fortunately, however, there are many close relatives to *Phoenix dactylifera* that will thrive around your home and grounds. Plant experts, especially the many fine foliage nurseries in Florida, offer a wide range of palms for your decorating pleasure.

Some fifty years ago, collections of palms of many types were common in halls and parlors as well as hotels. Now, interest in these exotic houseplants has revived, and, in response, firms like John's Dewkist Nurseries and Florida Growers are providing a wide choice of specimen plants to florists nationwide.

For home or office, palms are reliable container plants. They are surprisingly hardy and will thrive for years with a minimum of care.

Most directly representative of the date palm perhaps is the Pigmy Date Palm, *Phoenix roebelenii*. It is a compact plant with graceful, arching leaves which does exceptionally well in homes if kept warm and moist. For larger intensive decorating, the Christmas Palm, *Veitchia merrillii*, and the Solitaire Palm, *Ptychosperma elegans*, provide sturdiness and the impression of a full-sized palm tree on a small scale for use outdoors in southern climates, or as potted specimens indoors.

Palm trees do well as container plants because they have no deep taproots. In fact, they put up with crowding well, and if they have the semi-shade which they prefer, they can be kept for years. Tub or container culture also yields a bonus. Growing palms in containers to some degree controls their growth, keeping them within bounds for better appearance and less care. It is best to leave potted palms in their original containers as long as possible, even if the soil seems to be filled with roots. That helps restrict their size. Unlike other plants that become pot-bound, palms seem to take this condition in stride.

You should provide liquid fertilizer during the warmer season, which is the growing time for palms. Mix mild solutions of liquid fertilizer or organic materials, as directed on the fertilizer container. Watch the water needs of palms more closely than other plants. Palms prefer a plentiful supply of water, but they do not like wet feet. Soak the plants two or three times per week thoroughly, but be sure the excess water drains away into saucers to prevent their roots from standing in the water. If roots rot and die, the plants obviously cannot feed themselves, and they in turn will die.

When palms and ficus trees need repotting, use a rich loam, to which you have added generous amounts of peat moss and humus. That will improve the water-holding capacity as well as drainage. Be sure that you firm the old root ball tightly into the soil mixture when repotting. Otherwise, the water you add will channel itself around the root ball and your plant won't get the moisture that it needs.

You can select a variety of other palms, all close relatives of the palm of the scriptures. In fact, nurseries do offer the Arabian date palm, *Phoenix dactylifera*, as well as the general date palm, *Phoenix reclinata*. Many palms bear close resemblance to each other. If you are a purist, you may hold out until you can obtain the plant you really want. For most homes and for ease of care, the Pigmy or Miniature Date Palm, *P. roebelenii*, will work wonders as a replica closest to the true, tall date palms.

As you shop, keep in mind that the tree-type palms are distinctive. The fan palms, including *Chamaedorea elegans* and parlor palms, are another step removed from the date palm in appearance and growth habit.

POMEGRANATES

"A land of wheat, and barley, and vines, and fig trees, and pomegranates . . ." Deuteronomy 8:8

"And beneath upon the hem of it thou shalt make pomegranates of blue, and of purple, and of scarlet, round about the hem thereof; and bells of gold between them round about: A golden bell and a pomegranate, a

The sweet, refreshing fruit of the pomegranate was a favorite of the peoples of the Holy Land, as it is in areas of the Mediterranean where it grows so well today.

golden bell and a pomegranate, upon the hem of the robe round about."
Exodus 28:33–34

The pomegranate, mentioned in these verses and others, is quite clearly and correctly identified as a plant of the Bible: the pomegranate, *Punica granatum*. It was and is found growing in abundance in the land of the Bible. Although it usually is a small and bushlike plant, the pomegranate may, under suitable growing conditions, become a tree growing to a height of 20 to 30 feet. Its branches are opposite and alternating, and usually thorny. It isn't much of a tree to look at, but the showy bell-like flowers are appealing in red, yellow, or white, but mostly scarlet.

Pomegranates are easily identified by their fruit, which grows large and globular, the size of an orange or medium-sized apple. The hard rind is bright red with a yellowish cast when ripe. Pretty to look at, perhaps it is, but the pomegranate certainly ranks as one of the most difficult fruits to eat.

When the rains arrive in the flowering season in the land of the Bible, these trees begin to bud, as described in The Song of Solomon 7:12: "Let us get up early to the vineyards; let us see if the vine flourish, whether the tender grape appear, and the pomegranates bud forth: there will I give thee my loves." This lovely looking fruit, almost sparkling bright against blue skies, was compared with other beauty elsewhere in The Song of Solomon. In chapter 6:7, 11 you can read: "As a piece of pomegranate are thy temples within thy locks . . ." and "I went down into the garden of nuts to see the fruits of the valley, and to see whether the vine flourished, and the pomegranates budded."

In The Song of Solomon 4:3, 13 you will read: "Thy lips are like a thread of scarlet, and thy speech is comely: thy temples are like a piece of a pomegranate within thy locks" and "Thy plants are an orchard of pomegranates, with pleasant fruits."

It is believed by scholars that the bell-like flowers most probably were the model for the bells described so clearly in Exodus 28:33–34 and later in Exodus 39:24–26, which were embroidered on the temple robes. The fruit itself, it would seem, provided the model for the other decorations on the robes of those who were to minister to the people.

The pomegranate is a native of Asia, most likely from an area stretching from northern India to the Levant. It has been cultivated since prehistoric times. Today it is still common in the Holy Land as it is in Egypt and other countries bordering the Mediterranean Sea. In the scriptures it is placed in Egypt according to Numbers 20:5: "And wherefore have ye made us to come up out of Egypt, to bring us in unto this evil place? it is no place of seed, or of figs, or of vines, or of pomegranates; neither is there any water to drink."

In Deuteronomy 8:7–8, Moses reminds his people of God's goodness and the pomegranate appears again, together with other valued fruits and foods: "For the Lord thy God bringeth thee into a good land, a land of brooks of water, of fountains and depths that spring out of valleys and hills; A land of wheat, and barley, and vines, and fig trees, and pomegranates; a land of oil olive, and honey."

Although the pomegranate is a fairly productive tree, the fruit is, many will agree, nearly impossible to eat. For centuries the pulp has been used for making cooling drinks and sherbets, as well as eaten raw, but the vast quantities of seeds within the fruit make enjoyment of it somewhat of a chore.

Beneath the rind of the pomegranate is its sweet, juicy pulp in which large quantities of red seeds are imbedded. For those with patience, it is no doubt a most enjoyable, sweet fruit. In olden days, and in countries where it grows abundantly today, a spiced wine is made from pomegranate juice. Its soft seeds are also eaten and dried for use as a confection. As often done in ancient times, almost every part of this fruit is put to use, even the flowers, which yield a red dye, and the rinds, which have been used for tanning leather. "Waste not, want not" must have been a phrase well rooted in those days of the Bible.

As a tree of the hot weather in the Mediterranean zone, pomegranates respond to culture similar to that of the fig and olive. They prefer full sun and well-drained soil. Extra water is welcomed by these plants during the fruit-forming time to provide the sweetest taste as the fruit begins to swell and ripen in the sun.

The pomegranate has been favored by people other than Christians and Jews. Centuries after the children of Israel had wandered in the wilderness, longing for the cooling refreshment of the pomegranate, the prophet Mohammed advised his followers: "Eat the pomegranate, for it purges the system of envy and hatred." There is something about the juicy, subacid fruit which makes it particularly agreeable to the inhabitants of hot, dry regions. And it is in those arid areas in which the pomegranate attains its perfection.

This unusual plant was most likely introduced into the New World by the early Spanish colonists. For years before it had been grown in orchards of Spain and other hot Mediterranean countries. Today, the pomegranate is grown in the warmest parts of the United States, through Mexico, and in many areas of South America which have the appropriate climate for its best growth. Good fruit is produced only where high temperatures and dry atmosphere occur during the critical ripening period.

If you can obtain pomegranate plants, this fruit tree should be planted in deeply tilled, rather heavy loam. It is propagated by cuttings, which are rooted in open ground. In the United States, the three varieties of the pomegranate which are grown commercially are Wonderful, Paper Shell, and Spanish Ruby. You may also prefer to try growing dwarf forms in pots or tubs. They produce little fruit of value but are attractive plants when they bear their handsome, scarlet flowers.

CHAPTER X

TREES
OF THE BIBLE

"And ye shall take you on the first day the boughs of goodly trees, branches of palm trees, and the boughs of thick trees, and willows of the brook . . ." Leviticus 23:40

"And Joshua wrote these words in the book of the law of God, and took a great stone, and set it up there under an oak, that was by the sanctuary of the Lord." Joshua 24:26

"The glory of Lebanon shall come unto thee, the fir tree, the pine tree . . ." Isaiah 60:13

"The voice of the Lord breaketh the cedars; yea, the Lord breaketh the cedars of Lebanon." Psalms 29:5

"I have seen the wicked in great power, and spreading himself like a green bay tree." Psalms 37:35

In the earliest days of the Bible, the Promised Land was a land alive with foliage and trees, flowers and fruit. It was, as the scriptures so beautifully tell us, "a land of milk and honey." Forests covered the hillsides, rising up the slopes of the mountains. Stately trees, the cedars of Lebanon towering tall into the brilliant sky, massive oaks and pines, olive and fig and poplar and willow grew in their appointed places in the Holy Land.

Trees are as vital to mankind today as they were in Biblical times, when they provided the wood for homes and temples and churches. They yielded lumber for making furniture and fuel for fires. Flowering trees added their beauty to fields and forests as well as their abundant fruits for the nourishment of the people who lived among them. They provided the wood for carts and chariots, for rude shelters and gracious mansions. Now, as in Biblical times, mankind without trees would be poor indeed.

Trees, like all plants, are truly vital to life. Their leaves take in the carbon dioxide which we and all living creatures exhale. In fair return they provide us with oxygen, the oxygen which we breathe and must have to live. They give us shade and sustenance and all their other bounties as they did in the days of the Bible.

Through the ages, long before the birth of Christ and ever since, the Holy Land has been a corridor beset by troubles. The earliest tribes fought each other for possession of the land. Through the ages, warring factions and vast armies have battled in this narrow strip of land. In the process, forests were cut and burned, orchards chopped down, trees torn from the ground. There is no doubt that insects and natural diseases also played a part in altering the look of the land, the natural ecology of the land of the Bible. It is much different today from what it was in Biblical times, that land of milk and honey.

Despite mankind's unkindness to this land, it has survived, as have the people in it. The trees have stood as firmly as they could, withstanding all the ravages of time and wind and weather, pestilence and drought, and the thoughtlessness of man.

From the trees that survived, husbandmen propagated others, in groves and orchards and gardens to feed their families and their fellow man. Throughout the Bible, you will find hundreds of references to trees of all kinds. Despite the wantonness of many, there were others who understood the land and the trees that grew in it. They guarded them and respected them. As stewards of the land, many of the peoples who lived in the Holy Land from the beginning realized that the trees were part of their own vitality.

In the scriptures, you will find passages extolling trees for their virtues. The strength of oaks, the stateliness of cedars, the abundance of the olive and the fig are all vividly and often symbolically praised. Among the peoples of the Holy Land, many held trees in awe and reverence. They knew the importance of their trees to their life and the life of their children and their children's children.

Many learn from lessons of their ancestors. Others do not. Even today, modern bulldozers level trees so that houses, highways, and shopping centers can be built. Trees that took so long to establish their roots in those places are eliminated with the uncaring power of a dozer's blade.

Fortunately, today there is a realization rising across America. We are slowly coming to a greater appreciation of the wonder of God's world around us in the plants, the flowers and the trees that we should enjoy. A new mood seems to be growing in our land, an awareness and understanding that our environment depends on us to protect and improve it. With this awareness has come greater effort to preserve and conserve what we have among the plants and trees. All this is good.

As you read the scriptures, and this book, perhaps you too will find new hope for the future. As you plant the flowers, herbs, and vegetables, you can surround yourself with new beauty and enjoy tastier, more fruitful living. In this chapter, you can gain a new perspective on the largest plants that lived in the Holy Land, the trees. They were important to the people then as they are to us.

You can explore the numerous Biblical references to trees and expand your growing horizons too, planting some of these trees of the scriptures to enjoy in and around your own home and community. They will provide you, as they did the people of earliest times, with a sense of accomplishment, a renewed sense of pride in your land and the luscious fruits, the glorious blooms, the comforting shade, and the lovelier environment that all of us desire.

Trees are a thing of beauty. They can also be an asset to your home's value. Not only do well-landscaped grounds with attractive trees and flower gardens add to the beauty and appeal of a house, they also increase its real dollar value. Realtors have surveyed similar homes in various towns and found that those with attractive gardens graced by trees and flowers sell for several thousand dollars more than those without these growing assets. That's a fact.

You may plan to live in your house for many years to come. Statistics reveal that nearly 20 percent of Americans move each year. Put another way, one out of every five families will move this year. It seems we are a rather mobile nation. People move to accept new jobs or look for greater opportunity in other areas. Some upgrade their home environment as they earn more money or need a larger home when their family becomes larger. Others, in their retirement years, often look for a smaller home rather than maintain a rambling house that they no longer feel they need.

Whether you plan to move within the next five years or not, you can find it both enjoyable and practical to improve your present home plantscape now. For a small investment to buy the trees, plus your own planting efforts and tender loving care, they'll add many hundreds, if not thousands, of dollars to the value of your property.

Trees do much more, of course. They provide us with another dimension of God's beauty as they grow. Some add their flowering displays in their season. Others provide shade as well as privacy. Among plants of the Bible that trace their deep roots to the Holy Land, you have a wide choice of trees with which to make your personal environment more attractive and desirable.

In this chapter you'll find a selection of trees for bloom and trees for shade that you can grow easily in your own home ground. There are, of course, more trees than those in this book that are mentioned in the scriptures and native to the Holy Land. However, as with the other plants of the Bible in this book, only those are included that are readily available and are reasonably easy to grow in most parts of our country.

You will also find tree planting and care guidelines for each type of tree accompanying the descriptions. If you like nuts, you can grow the almond and the walnut. If you seek more color in your landscape, you may pick acacia, laurel, or tamarisk. Among the evergreens, you have a choice of cedar or pine; and mulberry, oak, poplar, or willow if you prefer deciduous trees.

Since trees will set their roots and grow handsomer while you plan and plant your flowers and your vegetables, they deserve priority in your list of things to do. Dig down in the good earth, improve the soil if it needs attention to give your trees better growing conditions. Add compost, peat moss, and manure as well for those trees that require more nourishment than your soil can provide.

Your trees get thirsty too, so be certain that you understand the needs of each type you plant and supply them the water and the fertilizer that they need. Fortunately, most trees, especially those species that flourished, or at least survived, in the sometimes harsh climate and environment of the Holy Land are reasonably easy to grow well.

Dig in soon. Trees of the Bible can be an important part of your life for many years to come.

ACACIA

"And rams' skins dyed red, and badgers' skins, and shittim wood . . ." Exodus 25:5

"And they shall make an ark of shittim wood: two cubits and a half shall be the length thereof, and a cubit and a half the breadth thereof, and a cubit and a half the height thereof." Exodus 25:10

"And thou shalt make staves of shittim wood, and overlay them with gold." Exodus 25:13

"Thou shalt also make a table of shittim wood . . . And thou shalt make the staves of shittim wood, and overlay them with gold, that the table may be borne with them." Exodus 25:23, 28

"I will plant in the wilderness the cedar, the shittah tree, and the myrtle, and the oil tree; I will set in the desert the fir tree, and the pine, and the box tree together." Isaiah 41:19

Search though you may through the scriptures of the Authorized Version of King James, you will not find the acacia tree mentioned by its name. That doesn't mean it is not a tree of the Bible. The shittah tree is mentioned once and shittim wood twenty-six different times. These references are always in connection with the ark of the tabernacle, its altar

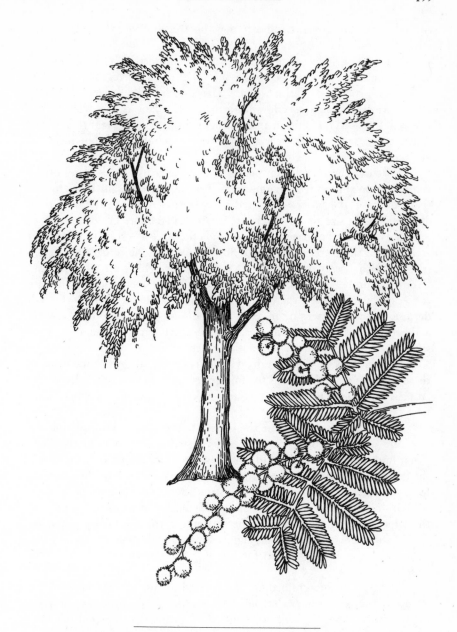

Acacia trees, growing where few other plants can survive, provided the people of the Holy Land with the shittim wood used so widely in ancient times.

and table, which seems to indicate that this tree and its wood were much respected and valued by the people of Israel.

You may well ask: if the acacia is not mentioned, why is it included in this book? Researching through the other versions of the Bible, the Douay uses the term "setim wood." That doesn't help much. However, in the Moffatt, Goodspeed, and Jastrow translations, as well as the Revised Version, the shittah tree and its shittim wood reveal themselves as the acacia tree and its acacia wood. Other scholars and botanists too have come to that same conclusion.

Actually, there are four species of the acacia tree in the land of the Bible, but no shittah tree, botanically. Tracing the roots of these trees to their source in the Holy Land, it is apparent that they are the only trees of considerable size in the desert areas. Acacia trees are basically trees of dry lands and barren places. Virtually no other tree can survive in the arid areas where these trees seemingly flourish. When one rereads the scriptures, it also becomes apparent that the times which are described in Exodus, Numbers, Deuteronomy, Joshua, Isaiah, Joel, and Micah were the years during which the children of Israel wandered through the wilderness. The only trees of any size common in the Arabian deserts were and are these drought-resistant acacia trees. Modern travelers say they seem to grow like weeds where nothing else takes root.

Today the wandering Arabs with their flocks still gather the wood of these trees for their fires, and the foliage serves as food for their livestock. Acacias have been important to the people of those desert regions since time began.

Acacia trees, admired for their persistence in such inhospitable terrain, are often shrubby, gnarled, and twisted with a windblown look. Their branches are covered with strong white spines in pairs that can tear the flesh. With better growing conditions, acacias can reach a height of 20 to 25 feet, but desert living seems to stunt them. Looking at one of these trees from a distance, its foliage of tiny bicompound leaves gives the tree the appearance of a fine fern. The flowers, borne in small groups, are yellow and the fruit is a slender, slightly curved pod. Acacia wood is hard, which makes it ideally suited for building the tabernacle as told in the scriptures.

Today, this fine-grained wood is prized for use in cabinet work. Its attractive orange-brown color polishes to a fine finish, producing appealing furniture. Thousands of years ago, the Egyptians also prized this hard wood to solidly bind shut their mummy coffins.

As you read the early scriptures, notably in Exodus, this wood is mentioned as the wood that must be used in making the ark and altar. In Exodus 37:1, 4, 10, 15, 25, and 28, which are partially quoted here, this valued wood is emphasized emphatically as the material to be used, seeming to bar the use of any other:

"And Bezaleel made the ark of shittim wood . . . And he made staves of shittim wood . . . And he made the table of shittim wood . . . And he made the staves of shittim wood . . . And he made the incense altar of shittim wood . . . And he made the staves of shittim wood. . . ."

Whether other trees, perhaps the desert date palms, were available or not, the shittim wood was the one that was to be employed.

Acacia trees are common in some areas of the Sinai, especially *Acacia seyal*. This species and another, *A. tortilis*, are, botanists agree, the acacia trees that are most likely meant in these many passages of the Bible. From the Dead Sea area southward, acacias can be found in abundance. They favor ravines or the wadis, with good reason. Although they can tolerate conditions few other trees can stand, acacias must have water at some time of the year. They obtain it from the rains that rush in brief, sporadic floods through these ravines before the water is swallowed by the desert sands.

As you read further in the scriptures in Numbers, Joel, and Joshua you will note that this word "shittim" is used to describe a place. In Numbers 25:1: "And Israel abode in Shittim, and the people began to commit whoredom with the daughters of Moab."

In Numbers 33:49 the scriptures tell us: "And they pitched by Jordan, from Beth-jesimoth even unto Abel-shittim in the plains of Moab." In Joshua 2:1 you find: "And Joshua the son of Nun sent out of Shittim two men to spy secretly," and in 3:1: "And Joshua rose early in the morning; and they removed from Shittim, and came to Jordan." Even in Micah 6:5, Shittim is a place: "O my people, remember now what Balak king of Moab consulted, and what Balaam the son of Beor answered him from Shittim unto Gilgal . . ."

In the days of the Bible it was common, as it is in many other parts of our world today, to name towns and areas after their most prominent landmark or attribute. So it seems to be with the acacia tree, the shittah tree that grows the shittim wood.

Not only was the acacia tree's wood valued highly, but it provided another value too. From this tree, by tapping its bark, the people obtained gum arabic for trade. That trade continues today as it has over the centuries from the earliest times in the land of the Bible. A less important but still useful product from acacia trees is the charcoal produced by burning them. The hard, densely grained wood produces excellent charcoal which is itself valued in the desert areas where this tree grows.

You can enjoy the wonders of this exotic tree, but not necessarily the exact species that grows in the Holy Land. However, other closely related species with their fine leaves will prosper in southern sections of the United States.

One of the chief advantages of acacias in warm areas is that they grow so rapidly you will have a feathery-leaved mature plant in very few years.

There are some 500 or more species of acacias known in warm regions of the world, particularly in Africa and around the Mediterranean Sea. The leaves are fine and compounded twice, which gives these trees a light, airy appearance, even lighter than the typical honey locust trees which grow in many areas of the United States. This type of leaf has its value in the hot regions, helping to reduce transpiration of moisture in the broiling sun. Most types have clusters of somewhat fragrant flowers which are followed by seeds in podlike fruits. Some species also have sharp spines on their branches.

Unfortunately, although these trees of the scriptures do grow rapidly, they have a rather short life span, twenty to thirty years. No matter; their advantages, the loose, feathery foliage and pleasant blooms, their ability to grow rapidly and withstand drought, make them useful in any hot, arid area.

Acacias are not fussy about soil needs. To start them well, stake the sapling until its roots are well anchored. Water it deeply but not frequently. The reason for thorough watering is to help it set deep roots rather than shallow ones which can dry out in droughts. Several species are available from nurseries, including *Acacia baileyana*, often called the Cootamundra Wattle, and *A. decurrens*, the Green Wattle, which is taller, more upright than the Bailey variety.

If you prefer to wander further afield from the look of the trees of the scriptures but stay within the acacia family, you might try the Weeping Myall, *A. pendula*. The narrow leaves are blue-gray in color and the branches actually cascade toward the ground. This tree will grow to 20 or 25 feet, but with a 12- to 15-foot spread, and bears yellow flowers sporadically in spring. If you have more room, *A. pruinosa* is one of the largest species and may reach 50 feet with a nicely dense, spreading crown.

Acacias are suited best for the warmer climates across the southern and western parts of the United States, so those of you who live farther north may have to forego the pleasure of these interesting trees in your home landscapes.

ALMOND

"And their father Israel said unto them, If it must be so now, do this; take of the best fruits in the land in your vessels, and carry down the man a present, a little balm, and a little honey, spices, and myrrh, nuts, and almonds . . ." Genesis 43:11

"Three bowls made after the fashion of almonds in one branch, a knop and a flower; and three bowls made like almonds in another branch, a knop and a flower: so throughout the six branches going out of the candlestick. And in the candlestick were four bowls made like almonds, his knops, and his flowers . . ." Exodus 37:19-20

Almonds were known early in Biblical times, not only for the beauty of their blooms, but also for their sweet and bitter nuts, their vital oil and the flavoring they provided to ancient peoples. Through the scriptures, from Genesis to Jeremiah, almond trees have inspired artists and provided food as well. Times haven't changed the almond tree. It is as much an asset now as it was when its roots grew in Biblical days in the Holy Land.

It is apparent from the scriptures in Genesis that almond trees were already growing in the Holy Land when Jacob sent for corn from Egypt, offering the present of the almonds and other nuts with balm and spices. It may be that almonds were not known or well known in Egypt then, about 1700 B.C., but reading through the scriptures, it appears they became a favorite there in later years. Much later, in the wilderness, almond blooms were used as models by the craftsmen-artists of that time for ornamenting golden candlesticks.

According to some Biblical scholars, the almond wood may also have been used by the chiefs of the tribes to make staffs as symbols of leadership. Quoting from Numbers 17:6–8: "And Moses spake unto the children of Israel, and every one of their princes gave him a rod apiece, for each prince one, according to their father's houses, even twelve rods: and the rod of Aaron was among their rods. And Moses laid up the rods before the Lord in the tabernacle of witness. And it came to pass, that on the morrow Moses went into the tabernacle of witness; and, behold, the rod of Aaron for the house of Levi was budded, and brought forth buds, and bloomed blossoms, and yielded almonds."

It is interesting to note that almond branches are famed for the speed with which they will break into premature bloom when placed in a glass of water in a warm place. You can try this yourself with a branch of an almond tree, or its relative, the apricot tree. Cut the branch in the early spring when the buds are full, bring it into the house and crush the base of the stem lightly. Then place the stem in warm water and within a few days you will see blooms open. In a small way, this is akin to the miracle of Aaron's rod.

The almond trees that are native to Persia and part of India spread westward during early caravan days of trade. It was a simple matter to carry the seeds in trade. From Palestine, they spread to Egypt and eventually to Europe and England, and arrived in America with the early settlers.

Almond tree leaves are long and somewhat oval, with a serrated margin. The fruit is like that of the apricot, round with a downy covering over its hard shell, but with none of the juicy flesh of the apricot. Only the almond seed itself is inside.

Almonds are tasty eating. If you live in the warmer areas where they will grow, you have a treat in store. Some of the hardier varieties now available can be grown in tubs in northern areas, then rolled on casters inside for decorative accents in your home.

Among botanists, there is still some debate whether almonds as we know them today are the almonds mentioned in the scriptures. The almond tree, *Prunus amygdalus,* has been widely grown in Asia and parts of North Africa for centuries. There are two types, those with sweet and those with bitter nuts. The kernels of the bitter almonds are about as inedible as apricot pits, but they contain great amounts of oil. When processed, this oil is used for flavoring extracts. Sweet almonds are used primarily as nuts for eating fresh or in cooking. Vegetables almondine is a gourmet's delight.

Almond trees are close relatives of apricot trees. For this reason, although scholars may disagree, it seems logical that they too were the trees of the scriptures. Whether that is fact or not, they are appealing as an attractive, fruitful tree for southern areas. The tree itself grows much like an apricot tree does. Its flowers are much larger and more expressive. The fruit, however, resembles an apricot's until nearly mature, when it turns dull gray, splits, and discharges its nut. If you have grown almonds, you'll realize that this fruit's skin is thin. Instead of fleshy, tasty fruit within, it is leathered and useless.

Although almonds are grown in scattered locations in many parts of the warmer regions of the Old World, the only commercial groves in America are in sheltered valleys of California. Like apricot trees, which can be nipped with frost that kills the buds and destroys a year's crop, early blooming almonds are sensitive to the cold. Some new varieties have been developed with somewhat greater hardiness, but it is best to plant them in protected areas, such as the sunny southern side of your home.

To many gardeners, almond blossoms are as valued as the nuts themselves. Flowers are pale pink or white and up to 2 inches across. The trees grow upright while young, but approaching maturity they spread into a dome shape, up to 30 feet tall.

For success with almonds, you must have good soil several feet deep that is well drained. Almonds are susceptible to root rot in wet soils and do poorly in shallow or sandy soils. You have a wide choice of varieties today. Nonpareil and Texas are two recommended, popular types. Once you plant the variety you buy, be sure to give your almond trees deep watering during their blooming and fruit-forming seasons. You can accomplish this easily once a month with a root-watering device that attaches to the end of your garden hose to deliver water easily and deeply underground through the hollow spike of the device.

As you pick and eat your almonds, think back to the miracle of Aaron's rod, the dry staff bursting into blossoms as described so beautifully in the scriptures. These trees are deeply rooted in our heritage.

CEDAR

"As the valleys are they spread forth, as gardens by the river's side, as the trees of lign aloes which the Lord hath planted, and as cedar trees beside the waters." Numbers 24:6

"Now therefore command thou that they hew me cedar trees out of Lebanon . . ." "So Hiram gave Solomon cedar trees . . ." I Kings 5:6 and 10

"The beams of our house are cedar, and our rafters of fir." The Song of Solomon 1:17

The roots of the cedars are firmly planted in the soil of the land of the Bible. Actually, the roots of the Hebrew word for cedar are derived from an Arabic word which signifies a "firmly rooted and strong tree." That they were!

At one time the mountains of Lebanon and other parts of Palestine were covered with these magnificent trees. Many soared 100 and more feet into the sky above the Holy Land. Their sturdy trunks, 6 to 8 feet in diameter, were deeply rooted in the land, awaiting their destiny as temples, homes, furniture, ships, and, of course, valuable cargoes of trade.

As these magnificent trees inspired awe and reverence among the children of Israel, they also attracted the attention of the rulers of the land as timber for use in trade.

"And Solomon sent to Huram the king of Tyre, saying, As thou didst deal with David my father, and didst send him cedars to build him an house to dwell therein, even so deal with me . . . Send me also cedar trees, fir trees, and algum trees, out of Lebanon . . . and, behold, my servants shall be with thy servants. Even to prepare me timber in abundance: for the house which I am about to build shall be wonderful great." II Chronicles 2:3, 8–9

Cedar was in great demand in the days of the Bible. With such an abundance of the tall trees, stout with many hundreds of board feet of lumber in each, it is understandable that these trees, with their aromatic wood, should be widely used. Cedar has an enduring quality to it. Like

The Cedars of Lebanon are among the best-known trees of the scriptures. You can grow them today in living tribute to their place in the Bible.

our redwood of the western forests, cedar resists decay and rot. Not only is the fine-grained, red-colored wood attractive, it has the added advantage of resistance to water. Small wonder that temples, homes, even chariots were made from it.

"And he built the walls of the house within with boards of cedar, both the floor of the house, and the walls of the cieling: and he covered them on the inside with wood, and covered the floor of the house with planks of fir" I Kings 6:15. "The beams of our house are cedar, and our rafters of fir" The Song of Solomon 1:17. "King Solomon made himself a chariot of the wood of Lebanon" The Song of Solomon 3:9.

It is apparent that cedar's fame was known throughout the land and far beyond, and highly respected for its multitude of uses. Cedar was extolled often in the scriptures, as in The Song of Solomon 5:15: ". . . his countenance is as Lebanon, excellent as the cedars" and in Isaiah 14:8: "Yea, the fir trees rejoice at thee, and the cedars of Lebanon, saying, Since thou art laid down, no feller is come up against us."

In allegory, parable, as well as symbolically, cedars find their way into passages of the Bible. The majesty of these gigantic trees is emphasized in Amos 2:9: "Yet destroyed I the Amorite before them, whose height was like the height of the cedars, and he was strong as the oaks . . ."

It comes as no surprise, therefore, that the forests of cedar were too well and perhaps unwisely used during the time of the Bible and through the centuries that followed. Wars claimed many more as rampaging armies cut these trees and burned forests. Today, in the Holy Land, only a few stately specimens stand tall, testimony to the glory of these ancient trees that graced the hills and mountains. Fortunately, these last remaining giants, like the redwoods in the national parks of our own west coast, are now protected to be preserved for the future.

The Cedars of Lebanon live on. Their heritage has been passed into the genetic pool of new generations of these fine old trees. You can obtain direct descendants for planting in your home landscape.

This most famous of cedars, *Cedrus libani*, is remarkably uncommon in American gardens. Perhaps that is the result of our being blessed with such sturdy native cedars in America. Inasmuch as the Cedar of Lebanon does grow well in most parts of the United States, it is time to begin cultivating it in more gardens of our country.

This species is somewhat slow in getting started, sometimes taking years before you notice that it has begun to change in size. However, don't despair. In reasonably fertile soil, and provided with adequate moisture, these trees can assume the stature that they attained in the scriptures. Cedars of Lebanon are, and can be for you, truly stately trees of the Bible.

You can grow them or related species, *Cedrus atlantica* or *C. deodara*, best where they have room to spread out naturally. Wild native cedars in eastern states tend to be tall spindly looking trees. Cedars of Lebanon and close relatives are far from that. Needles of cedar trees, by the way, are grouped together in somewhat tufted clusters. They also bear their cones upright on the branches.

The first few years you'll notice that your Lebanon cedars tend to form heavy limbs on the lowest portions of the trunk. In time, these may curve upward to form what appear to be secondary trunks. This is a natural habit which at maturity creates a much wider tree with a somewhat flat-headed crown. You can prune some of these low branches away if you prefer, to make room for lawn chairs beneath these boughs.

Cedars of Lebanon will probably surprise you as they grow. Changing from the narrow, pyramidal appearance of their youth, they will become more open, irregular, and magnificent as the years go by.

At Christmas time they offer you another reward. You may clip some branches to decorate your home for the holidays, some boughs of the Cedars of Lebanon so deeply rooted in the land of the Bible.

LAUREL

"I have seen the wicked in great power, and spreading himself like a green bay tree." Psalms 37:35

It may seem strange that such renowned translators as Drs. Goodspeed and Moffatt regard this tree as the Cedar of Lebanon, *Cedrus libani*. Perhaps they base that determination on faulty old texts which they researched during their work. Today, nearly all botanical authorities regard it as the Sweet Bay Laurel, *Laurus nobilis*. Logic tells us that is the most likely identification for the green bay tree of the scriptures in Psalms.

The Sweet Bay Laurel is a native of the land of the Bible. This evergreen tree grows in woods and thickets from coastal areas to the foothills of mountainous regions of the Holy Land. It has a rather compact growth habit, with oblong to somewhat lanceolate leaves. They are thick in texture and dark, glossy green. The fruits of this tree are berry-like, about the size of a tiny grape.

You may not recognize the tree, but no doubt you have used its leaves. This bay leaf has a pleasantly spicy fragrance and is widely used around the world in cooking. You may have used dried bay leaves in

Since the earliest times and through the Greek and Roman periods, laurel trees provided sprigs for wreaths to honor prominent citizens.

making such prosaic dishes as spaghetti sauce. It adds its rich bouquet just as well to other, more exotic gourmet dishes.

The laurel tree offers other wonders. It provided in those days of the scriptures the leaves that yielded a fragrant oil, called oil of bay. The fruits, roots, and bark, as well as the leaves, have been used medicinally for centuries.

Through the Greek and Roman periods, laurel wreaths were a sign of nobility and honor. From those times comes the Latin word "baccalaureate," which means, literally, laurel berries. The wearing of bay leaf wreaths by poets and scholars signifies a mark of distinction to this day as the word implies when receiving that degree from colleges and universities. These wreaths in ancient times were made of laurel branches and sprigs from this tree, the Sweet Bay Laurel, the "noble laurel," *Laurus nobilis,* as its name implies.

You can grow these plants in your gardens in southern areas where winters do not drive temperatures much below freezing. The laurel will eventually reach tree height, perhaps as high as 40 feet, but these plants are notorious for their slow growth. They may seem to take forever to grow beyond their shrub stage. That slow growth is a natural characteristic of the sweet bay tree. Because it grows so slowly it is a good plant for hedges, forming compact borders which are especially attractive. The polished green foliage is its best asset. The flowers are small. creamy yellow and, like the tiny purple berries, not especially attractive.

Laurels grow in most types of well-drained soil. You should water newly set plants until they become well rooted. After that, laurels as hedges or trees need little care. You can prune them back into formal hedge shapes or treat them as specimen plants if you wish.

The shiny, leathery leaves of the Sweet Bay Laurel are another bonus you receive from this unusual and functional plant of the Bible.

MULBERRY

"I clothed thee also with broidered work, and shod thee with badgers' skin, and I girded thee about with fine linen, and I covered thee with silk." "Thus wast thou decked with gold and silver; and thy raiment was of fine linen, and silk, and broidered work; thou didst eat fine flour, and honey, and oil: and thou wast exceeding beautiful, and thou didst prosper into a kingdom." Ezekiel 16:10 and 13

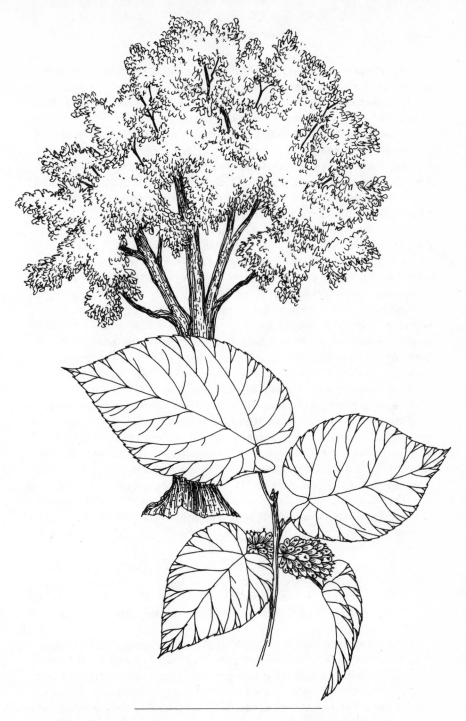

The mulberry tree and its bushy relatives are rooted far back in Biblical days.

Although botanists say that the Black Mulberry, *Morus nigra*, is most likely a native of northern Persia, now Iran, it has been cultivated for untold generations in the Holy Land. It is loved for its delicious fruit, sweet and juicy, which it produces in abundance.

You may wonder why the first quotation from the scriptures of Ezekiel belongs here since mulberries are not mentioned. In the Orient, the Chinese and Indian species of mulberry trees have long been grown to feed the silkworm. We may today be accustomed to synthetic fiber in the clothes we wear, but in ancient times, wool from sheep, linen from flax, and silk from the lowly silkworm feeding on its favorite food, the mulberry leaves, were most common materials from which the people made their garments.

In fact, today in Syria as well as in Palestine and in some neighboring countries, the Chinese and Indian mulberry species, *Morus alba*, is widely cultivated to grow silkworms.

Biblical scholars believe that Ezekiel became aware of silk during his captivity in Babylon. His is the first mention of silk by name in the scriptures, although some other scholars feel that Solomon, who lived some four hundred years before Ezekiel, may have known it. Ancient Greeks grew mulberries, as did the Romans, for the silk they could obtain from the busy silkworms feeding on the plant's leaves.

In Revelation 18:12 you will realize that silk was one of the prized products of Babylon: "The merchandise of gold, and silver, and precious stones, and of pearls, and fine linen, and purple, and silk, and scarlet, and all thyine wood, and all manner vessels of ivory, and all manner vessels of most precious wood, and of brass, and iron, and marble . . ."

Silk may have been a valued cloth during the days of the children of Israel and long after, but silk and the mulberry tree date back even further in time. The mulberry tree was being carefully tended for the prized silkworms in China approximately 4,000 B.C. according to very ancient records.

The black mulberry of the Holy Land is a rather low-growing and stiffly branched tree. It matures from 25 to 30 feet with a stout trunk. It is deciduous, with lobed and unlobed oval leaves. The purple-black fruit can be eaten fresh, made into preserves, crushed for wine, and even used in producing dyes.

Mulberry trees have been mentioned in the scriptures, as you realize, but you may wonder why they are not more widely grown. For one thing, they produce their mulberries, which attract many types of birds. Droppings from the birds stain outdoor furniture, car finishes, and clothes.

In the Orient and parts of the Middle East, mulberry trees are still grown widely as host to the valuable silkworm. A fast-growing tree, like poplar and willow, the mulberry tree can be grown in warmer areas of the

United States. Since many people object to the stains caused by the berries, a nonfruiting variety has become more popular. This White Mulberry, *Morus alba*, is extremely fast-growing as a shade tree and does exceptionally well where other trees don't thrive, in hot areas and those with alkaline soils. Once established, these trees may grow 35 to 50 feet or more, with a round-topped shape. The fruit-bearing species grow somewhat taller.

Mulberry trees and grafted weeping varieties are available. Fruitless and Stribling are two fine varieties for home grounds. Leaves may be 6 inches long and 3 to 4 inches wide, turning bright yellow in the fall. A weeping type, *M. alba 'pendula'*, is fun to grow since its branches grow from an upright stump and flow in a cascading pattern toward the ground. These are satisfactory for more northern areas and bear berries that attract the birds. With proper pruning, you can keep these in a moderate size for small plots.

When you plant a mulberry sapling, give it lots of water for the first year or two until it becomes firmly rooted. Then, normal rainfall will usually be sufficient for its growth and prosperity.

OAK

"And they gave unto Jacob all the strange gods which were in their hand, and all their earrings which were in their ears; and Jacob hid them under the oak which was by Shechem." Genesis 35:4

"But Deborah Rebekah's nurse died, and she was buried beneath Beth-el under an oak: and the name of it was called Allon-bachuth." Genesis 35:8

"And they made upon the hems of the robe pomegranates of blue, and purple, and scarlet, and twined linen." Exodus 39:24

"And Joshua wrote these words in the book of the law of God, and took a great stone, and set it up there under an oak, that was by the sanctuary of the Lord." Joshua 24:26

In Israel, in northern Africa, in Asia, Europe, and North America as well, small and mighty oaks have been growing for thousands of years. Although these trees are mostly native to temperate regions of the world, they also thrive in higher elevations in warmer areas. Certain species, their trunks gnarled with age, still are firmly rooted in the land of the Bible.

Oaks are in the genus *Quercus* but the many species differ in their size, growth habits, and the shape of their leaves. When most of us think of oaks, we think of tall, massive trees, often overlooking the scraggly scrub oaks common to sandy, pine-covered coastal plains. In southern New Jersey, for example, you can find a wide variety of oaks. In the northern areas tall, rugged trees grow well. In the sandy pine barrens, amid the pine trees, the stubby, wind-blown scrub oaks are prominent.

So it is in the land of the Bible. Botanists and Biblical scholars have studied the scriptures long and hard, and now identify four types of oaks that most likely were included among mentions of the oaks and "groves" of trees. These include *Quercus ilex*, *Q. macrolepis*, *Q. coccifera*, and *Q. lusitanica*.

There is no doubt that oaks were to be found in the Holy Land in Biblical times since they are native, as other species are throughout the world. Tracing their traits, the four species mentioned above are the most likely to have been in Palestine during the earliest times, although other types have been brought to modern Israel through the years.

Oaks, with their tight grain and dense wood, have always been valued for use in construction and furniture building. In homes today, oak floors are a mark of distinction and quality. The sturdy oak timbers of old barns built in the pioneer days of this country have endured the test of time where lesser woods have buckled and warped.

Oaks were as reliable for building in the days of the Bible, but obviously not as well thought of as the cedars which are mentioned so many times in speaking of building temples and great homes.

In ancient Palestine, some of the oaks were the smaller, somewhat scrubby types, growing 8 to 20 feet tall in the moutainous regions from Lebanon through what is now modern Israel. These lower growing species branch from their base and bear many leaves which are small, smooth, and evergreen with oval shape. They still produce acorns bountifully. In other areas, the oaks were more massive, putting down permanent roots that supported their more lofty stature. These were the trees often mentioned as "groves" in the scriptures, according to many scholars, as in I Kings 14:23: "For they also built them high places, and images, and groves, on every hill, and under every green tree."

You also will find many mentions of "scarlet" in the scriptures, which are in reality a way of referring to oak trees. The art of producing dyes was known to people at a very early time in history. Scarlet thread, for example, was employed at the time of Zarah's birth, 1700 B.C. Dyes were obtained from a variety of plants, flowers as well as trees. Tiny scale-like insects which infest young oaks produce a scarlet dye. It is this colored dye that is associated with the insect and the oak trees, scholars believe, which resulted in the references to "scarlet" in the scriptures.

Stately giant oaks are mentioned in the scriptures and were often picked as a site for burial of leaders. This Holly Oak is similar to those that grow in the Holy Land.

Reading through the scriptures more thoroughly, you'll find the oak mentioned first in Genesis. This, modern botanists believe, is the Holm or Holly Oak, *Quercus ilex*. It is common from Syria to Judea and flourishes in seacoast regions elsewhere. A most tolerant tree, it thrives in poor soils, even when beaten by winds and salt air. For this reason, the Holm Oak is a reliable plant for similar situations in the United States.

In the scriptural references to "thick" trees, some authorities feel that the trees indicated are oaks, while another school of thought favors pines with their thick clusters of needles, rather than "thick" meaning the girth of the tree. You may interpret these passages as you wish, of course, but the fact remains that oaks were and are common trees long rooted in the Holy Land.

The taller oaks appear to have been valued and revered for their substantial qualities by the peoples of the Holy Land. They are noted as the symbol of strength, as oak is still valued today. Today oak is still used in small boat construction, and for other nautical purposes. You can see from Ezekiel 27:6 that the oak was useful also in Biblical times: "Of the oaks of Bashan have they made thine oars; the company of the Ashurites have made thy benches of ivory, brought out of the isles of Chittim."

The strength of oaks is used symbolically in Amos 2:9: "Yet destroyed I the Amorite before them, whose height was like the height of the cedars, and he was strong as the oaks; yet I destroyed his fruit from above, and his roots from beneath." Oaks also were mentioned, as were groves, as places where worship was conducted by heathens as in Hosea 4:13: "They sacrifice upon the tops of the mountains, and burn incense upon the hills, under oaks and poplars and elms, because the shadow thereof is good . . ."

The tall oaks mentioned by Isaiah 2:13, "And upon all the cedars of Lebanon, that are high and lifted up, and upon all the oaks of Bashan," again refer, it seems, to trees of strength and stature. In other passages, scholars note that these trees were often chosen as the markers for the graves of strong men and leaders.

You may also find references to cypress as you read the Bible. Although some scholars believe this must be the cypress tree itself, there is an oak that is called Cyprus Oak, *Quercus lusitanica*. That perhaps accounts for the misinterpretation.

It is true, as some have pointed out, that the climate of Israel is generally too warm for most types of oak to flourish. However, temperatures on the hills and mountains are suitable, as those oaks growing there today readily prove. Being long-lived trees, some of the oaks that have survived the many centuries of alternating war and peace in Palestine stand today. Their great girth is silent testimony to their ability not just to survive but to thrive. Since oaks do thrive under a variety of climatic and soil conditions, you too can grow them where you live. You have, of course, a wide choice from among the different American species, as well as those more closely resembling oaks of the Holy Land. All, in fact, are related in the genus *Quercus*, so pick which ones work best for you.

Mighty oaks from little acorns grow. You can start your own oak tree that way, of course, planting an acorn in a pot until it sprouts, then growing it to a sapling a few more years until it is ready for transplanting to a permanent place in your home landscape. More practically, it is sensible to obtain a sturdy young tree from a local nursery.

Oaks are mighty trees, generally speaking. Some, of course, are more slender and may fit into limited space, if that is a concern. Most of the American oaks native to our country look different from the oaks mentioned in the Bible. The Live Oaks of our South are massive. Pin Oaks of northern, swampy areas are upright and somewhat pyramidal in shape. The oak of the scriptures is best exemplified by the Holly or Holm Oak, *Quercus ilex*, which you can grow successfully in the United States in most areas.

Since many types are available and perhaps more suitable for your purposes, you can select the species which suits you best. Local nurseries can provide those that thrive in your locality. However, the Holly or Holm Oak is closer to the appearance of the oak mentioned in the Bible. This tree has lightly toothed leaves 3 inches long and about 1 inch wide, glossy and dark green with undersides somewhat yellow or silvery. This tree tends to be dome-shaped and rather dense, and may grow from 40 to 60 feet tall.

In keeping with their heritage from their ancestors of the Holy Land, Holly Oaks are tolerant of somewhat poor soils and the salt air and wind of seaside areas.

Whichever oak you decide to grow, and there are many in this family, you can in time enjoy its shade as those in Bible times did so many thousands of years ago.

PINE

"And ye shall take you on the first day the boughs of goodly trees, branches of palm trees, and the boughs of thick trees, and willows of the brook; and ye shall rejoice before the Lord your God seven days." Leviticus 23:40

"The glory of Lebanon shall come unto thee, the fir tree, the pine tree, and the box together, to beautify the place of my sanctuary; and I will make the place of my feet glorious." Isaiah 60:13

Great pines still grow in the Holy Land, and you can duplicate their appearance with pines available for your home landscapes such as this Aleppo Pine.

Depending on which translation of the Bible you care to read, you will find references to pine trees. However, you may compare the same chapters and verses in different versions of the Bible and find that in one the pine is called an "elm" while in another version it has changed to a "plane" tree. This confusion can be traced in part to the lack of botanical knowledge on the part of some of the early translators of the scriptures. It can also be attributed to confusion about the type of tree that can grow in the climate of the Holy Land. Some scholars tended to equate a foreign word for a tree with a plant with which they were familiar in their own native land.

No matter how the errors of identification occurred, one thing is clear: there were and are pine trees in the land of the Bible, as there were and are in many areas surrounding it.

The references to "thick" trees also no doubt led to some misinterpretations. It would be logical for some translators to assume "thick" might refer to the breadth of the trunk of a tree, such as a stout oak. The passages in Leviticus certainly do seem to say "thick" trees. And, in Nehemiah 8:15 you also find that reference: " . . . fetch olive branches, and pine branches, and myrtle branches, and palm branches, and branches of thick trees . . ."

Again in Psalms 74:5–6, you see the term: "A man was famous according as he had lifted up axes upon the thick trees. But now they break down the carved work thereof at once with axes and hammers."

Biblical scholars have puzzled over these passages and words for years. Fir, pine, thick, cedar are seemingly applied rather loosely in some parts of the scriptures, while they seem more definitive in others. Remembering that those who wrote the Bible were not men of science, perhaps we can understand their loose use of names for trees. Even today, it is difficult for many of us to distinguish some pines from firs or spruces, or other evergreens for that matter.

However, despite the variations in the different translations and versions of the Bible, it seems obvious that pines are truly a tree of the Holy Land. They grow there today and botanists agree that some species can trace their heritage to the days of the Bible. According to botanists, including translators who studied botanical references, the most likely pines are the Brutian Pine, *Pinus halepensis 'brutia'*, and Aleppo Pine, *Pinus halepensis*. If you interpret "thick" trees to mean those that appear thick with needles as a pine does, some of the passages may seem clearer.

Pinus halepensis 'brutia', commonly called the Brutian Pine, is known to have existed in the mountains of northern Palestine and Lebanon. According to the scriptures, however, those areas were not occupied by the children of Israel during the time of Isaiah and Nehemiah. Brutian Pines did grow, however, in some parts of Palestine where the people were led by Joshua. That makes it quite conceivable that it was indeed pines that the faithful were told to use for making booths for the Feast of the Tabernacles.

Since most authorities agree that such pines did grow in the Holy Land, we accept that determination. The Brutian Pine grows to a height of 30 feet with somewhat whorled branches. The needle leaves are borne in clusters of two which are thicker than the Aleppo Pine, which also still is found in Israel today, a much larger, more stately tree. In Isaiah 60:13 the pine is again mentioned, along with others that are the "glory of Lebanon," which, "shall come unto thee, the fir tree, the pine tree, and the box together, to beautify the place of my sanctuary." It seems appropriate that the fullness of the evergreen pine would merit such description.

In other passages, in Samuel, Kings, and Chronicles, the fir tree is mentioned. You can read in II Samuel 6:5: "And David and all the house of Israel played before the Lord on all manner of instruments made of fir wood, even on harps, and on psalteries, and on timbrels, and on cornets, and on cymbals." In II Chronicles 2:8, firs are again specified: "Send me also cedar trees, fir trees, and algum trees, out of Lebanon . . ."

There are many references to cedar and fir in the book of Isaiah, and that tree appears again in Ezekiel, Nahum, and Zechariah. In keeping with most authorities, we think that the fir tree mentioned is another type of pine, most likely *Pinus halepensis*, the Aleppo Pine. This tree may be somewhat scraggly at times, but it can soar to 50 or 60 feet in height. The Aleppo Pine also is a native of the Mediterranean area and is found most abundantly on the hills in Palestine and Lebanon.

Since this tree is mentioned in the same breath as the stately cedars, it is likely that it was often used for building in the days of the Bible. Not only was it used for homes and temples, but in boat building as well, as indicated in Ezekiel 27:5: "They have made all thy ship boards of fir trees of Senir . . ." When you read Hosea 14:8, you also find reference to a "green fir tree" in several versions of the Bible, including the Authorized and Douay.

The Aleppo Pine, *Pinus halepensis*, may not be the most handsome tree to your eye, but it can thrive in poor soil, arid conditions, and high heat while being abused by wind and salt air. No wonder that its growth is often so irregular. Its green needles, borne 2 per cluster, have a heavy waxy coating to resist evaporation of moisture under the very harsh conditions in which this hardy tree grows. Because they persist in adverse conditions, it follows that Aleppo Pines do well in almost any type of soil, especially along coastlines in southern areas.

Since it has been well established that tall pines did grow in the Holy Land, pine trees deserve a place in your landscape. Unfortunately, there has been little demand for these two species of pine trees in this country. We have many other fine native pines. You can select some of those spe-

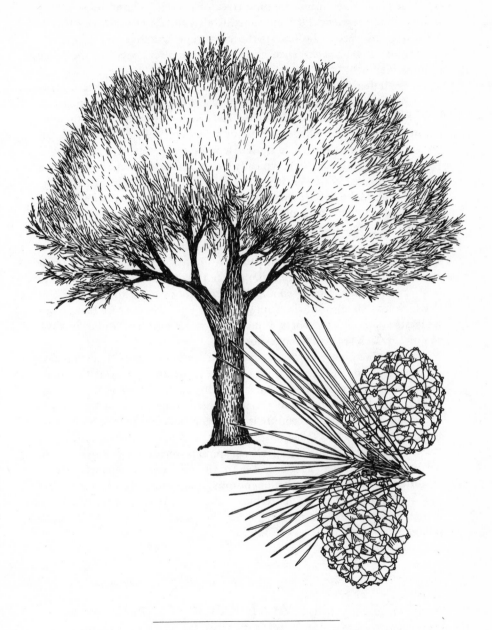

A tree that is quite similar to those grown in the Holy Land is the Italian Stone Pine. It prefers the warmer climates.

cies that closely resemble the pine trees of the Holy Land or select an
Aleppo Pine to grow. That choice is up to you. *Pinus pinea*, the Italian
Stone Pine, seems to be one of the most representative. The Japanese
Black Pine, *Pinus thunbergiana*, also bears some resemblance to the pines
of Palestine.

The Italian Stone Pine prefers the warm weather of southern areas
and will not survive in northern latitudes. One of the easiest pines to
identify, it soars higher than many and has a distinctive, flat-topped,
umbrella shape. The foliage is dense on wide-angled branches. Although
growth is slow, this pine, like those Aleppo Pines of the Holy Land, will
grow 60 to 80 feet tall, with typically heavy limbs. Like the pines of the
Bible, its needles are in clusters of 2 and are about 6 to 8 inches long.
Once established in sandy loam soil, which it prefers, it will endure heat
and even drought. This giant and stately pine tree performs best near the
sea, as its relatives do in the land of the Bible.

The Japanese Black Pine also begins life slowly, but eventually
speeds up its growth and may mature to 100 feet tall. It too has an open
growth but more of an oval shape, unlike the spreading top of the stone
pine. The Japanese pine, *Pinus thunbergiana*, does well in zones 5 through
8, which include most of the United States. This also is a tree that favors
seacoasts and seems to benefit from salt spray. It also has needles in clusters of 2, which are 4 to 5 inches long.

Both these pines seem well suited as representatives of Biblical pines.
You may, of course, elect to grow native pines as replicas of the pines
mentioned in the scriptures. Your local nursery can advise you which do
well in your locale. Fortunately, once planted, pines fend for themselves
quite well with little care, and their evergreen foliage is part of your landscape all year long.

Remember that pine needles do drop, to be replaced by others. These
fallen needles create an acid condition in the soil beneath pine trees that's
ideal for growing azaleas, rhododendrons, and other acid-loving shrubs.
But, if you wish to grow vegetables and other plants that need a more
neutral soil, plant those gardens elsewhere, unless you plan to sweeten
soil regularly with additions of lime.

POPLAR

"They sacrifice upon the tops of the mountains, and burn incense upon
the hills, under oaks and poplars and elms, because the shadow thereof is
good . . ." Hosea 4:13

Pictures of modern Israel reveal long rows of tall poplars along some roads and in groves wherever there is water for their ever-thirsty roots to drink. Poplars, *Populas alba*, and probably several other species have been growing in the Holy Land since the earliest times. Biblical scholars and botanists alike agree to that fact.

These tall trees, producing welcome shade, also have found their admirers in many other parts of the world. In France and Italy especially, poplars border highways and farm fields. They serve as windbreaks and are frequently used to provide their dense shade on dwellings.

Poplars are first mentioned in the book of Genesis 30:37–38: "And Jacob took him rods of green poplar, and of the hazel and chestnut tree; and pilled white strakes in them, and made the white appear which was in the rods. And he set the rods which he had pilled before the flocks in the gutters in the watering troughs when the flocks came to drink, that they should conceive when they came to drink."

There remains some debate whether the poplar in this passage is the *Populus alba*, the White or Silver-leaved Poplar. That really matters little, since all evidence points to the fact that these trees are indigenous to the Holy Land, and most Biblical and botanical scholars now accept the poplar that grows there as the poplar of the scriptures. Some authorities seem to challenge for the sake of debate at times. Those who believe the poplar was the storax tree fail to realize that this is a scrubby plant with short branches, certainly not what seems to be pictured in the scriptural references to the poplar tree.

Actually, white poplar has been common in moist areas and along streams or waterways in Palestine, Lebanon, Syria, and even in Sinai for as long as records have been kept.

These deciduous trees bear shiny green leaves with white and woolly undersides. They vary in shape from rounded to heart-shaped and oval. Young buds are covered with a resinous gum and smell somewhat like balsam when first rising in the spring. Some experts believe that this resin formed by the poplar buds may have been burned by Ephraim in the groves of poplars. Most scholars, however, indicate that the incense used primarily in the Holy Land was frankincense.

You can grow the white poplar if you wish. In recent centuries it has gained fame as the so-called Lombardy poplar of Europe. It may have a new name, but there is no doubt that this tree's roots belong in the land of the Bible.

Poplars are in the *Salicaceae* family, which also includes the willows. The large Lombardy poplar so fashionable in parts of Europe and along some of the old roads there that date back to Roman times are gigantic trees, often attaining a height of 100 feet. The *Populus nigra* with its '*italica*' variation is rather remarkable for its fast growth, tall size, and almost cypress-like shape. In this respect, judging from the words used in the

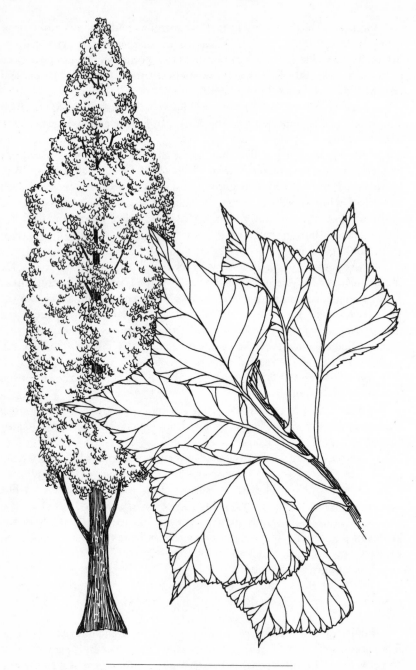

Poplar trees stand tall and impressive even today as they did in Biblical times in the Holy Land.

scriptures, this tree may well be the one mentioned as a cypress in various passages. It also could have been a cypress oak, of course.

The native land of this dramatic poplar is probably Persia, now Iran. Other botanists give its native soil as somewhere in the eastern countries around the Mediterranean Sea, from which it was, no doubt, brought to Europe centuries ago. Poplars enjoy moist soil and can be found flourishing along streams and river banks as well as drainage ditches where ample water is available to support their rapid growth.

Another characteristic of this type of tree is the formation of catkins, which are the seed-producing parts of the tree. Leaves are oval in shape and give the tree an airy look. In the United States this tree does well in moist, fairly fertile soils. A close relative commonly found throughout the western states is the cottonwood tree, which also occurs in eastern states in a slightly different form. American poplars also produce lumber for building homes, as did the poplars of the Bible land.

Poplars are valuable for their fast growth, serve well as tall windbreaks, or to screen your land from views of unsightly areas. The columnar shape lets you plant several in a row as a backdrop for other lower-growing trees and shrubs. In the fall, the leaves turn golden yellow, a brilliant sight among other autumn foliage of reds and golds.

You will find this tree grows well and rapidly with little care. It is susceptible to a canker disease that can harm the tops of the trees. Modern pesticides do help solve the problem.

Another type you might consider in the same family is the Quaking Aspen, *Populus tremuloides*. It is in reality a tree of more northern latitudes, but since it is related, it may fit into your perspective. The fluttering leaves, even when no breeze seems to be stirring, are a fascinating and welcome sight, since they give you the mental impression of a cooling breeze in hot summer weather. These too prefer moist soil.

Because poplars send feeding roots in search of moisture, never plant them near water lines, and especially never near septic tanks. Their roots can reach into the smallest crevice and clog the pipes of your drainage field rather rapidly.

TAMARISK

"And the water was spent in the bottle, and she cast the child under one of the shrubs." Genesis 21:15

"When Saul heard that David was discovered, and the men that were with him, (now Saul abode in Gibeah under a tree in Ramah, having his

spear in his hand, and all his servants were standing about him;) . . ." I
Samuel 22:6

You may not find any references to the tamarisk or tamarix tree in
the scriptures of the Authorized Version of King James. The passages
above are vague at best if you are looking for definitive identification.
However, in the versions translated by Drs. Goodspeed and Moffatt as
well as in the Jastrow version, you will find the words "shrubs" and "tree"
now are "tamarisk" and "the tamarisk tree." Modern authorities, including
botanists, agree that this identification is correct.

In areas of dry land, the tamarisk tree abides because it has a tolerance for adverse conditions that few trees have. Actually, it is more a
shrub, struggling to survive in these harsh regions where few others
could. In the passages from Genesis describing the wandering of Hagar
with her outcast child, the areas described are desert. There, the stunted,
barely surviving shrubs of tamarisk grow. This plant, *Tamarix aphylla*,
and another species, *Tamarix tetrandra*, may still be found in numbers in
those deserts.

Depending where they grow, the tamarisk trees may be truly small
trees, if they find sufficient moisture to nurture their taller growth. In
most dry areas, they are shrubs, often growing in dense masses. As trees,
they are more graceful with feathery branches bearing the tiniest leaves.
In spring, these trees burst into bloom, their flower spikes covered with
pink blossoms to give the appearance of a pink ball of bloom.

Since the tamarisk is so abundant in the harsher areas where plant
life of any kind is difficult at best, it would seem strange not to find some
reference to such common plants in the scriptures. In I Samuel 31:12–13
you find the story of the burial of the bodies of Saul and Jonathan. Some
scholars, pointing out that oak trees were more likely the resting places
for great men, believe that oaks may have been meant. But in that forbidding area which is mentioned by name the tamarisk tree appears to be
most common: "All the valiant men arose, and went all night, and took
the body of Saul and the bodies of his sons from the wall of Beth-shan,
and came to Jabesh, and burnt them there. And they took their bones,
and buried them under a tree at Jabesh, and fasted seven days." To this
day, thickets of tamarisk can be found near the site of Jabesh-Gilead.
Many scholars believe that the Hebrew word "eshel" from early texts of
this burial story should be translated as "tamarisk."

If you live in a harsh dry environment and have trouble growing
plants, perhaps these trees, or their shrub forms, will add color and their
Biblical meaning to your land.

The tamarix, called more commonly tamarisk, is actually not much
of a tree since it is low-growing and appears to be more of a shrub. However, its value has been recognized for planting where few other trees will

Tamarisk trees thrive in arid conditions. You can grow these relatives today as beautifully flowering bush-like trees in dry areas of the United States.

grow. It survives well in salty areas, by seashores, and on what might be considered to be almost sterile locations in hot, and semitemperate locales. The long, slender branches bear many small leaves, and their surfaces are tough, resisting loss of moisture.

It is a rather attractive shrub-tree in its blooming period, bearing clusters of tiny flowers. A chief value of this tree is its ability to grow surprisingly well in alkaline, sandy soils in dry climates. Few other plants can flourish in such adverse conditions. In this ability alone it is worth considering for use in desert areas of our country. The Athel tree, *Tamarix aphylla*, is a native of Asia and has been introduced to America where it is favored for windbreaks in hot, arid regions such as Southern California.

Several nurseries offer Tamarisk trees and shrub varieties. Pink Cascade, *T. pentandra*, is a beautiful shrub with masses of cascading panicles of flesh-pink blooms. The foliage is rich blue-green, and trusses are borne above this foliage from July into September. This shrub does well in seashore areas because, like its larger relatives, it is especially tolerant to salty air and wind. Summer Glow, *T. pentandra*, is noted for its lacy appearance with silvery, blue-green foliage which is topped by rose-pink flower spikes all summer long. The flowers serve well for cutting. You can prune this tamarisk to keep it low, but it normally matures to only 8 feet in height.

The tamarisk should be carefully planted, and the shrub pruned back severely during transplanting, as well as every spring. This is important to encourage the necessary new growth that will keep the plant producing flowers as it should.

Tamarisks also fill a need, not only for blooming beauty in dry, salt-air areas, but for their significance as representative plants of the scriptures.

The tree varieties grow rather rapidly to a height of 30 feet and have a wide, billowing, almost feathery appearance in their foliage. You can grow these only in warmer areas in the United States, but be aware that they have a wide-ranging root system which competes aggressively with other plants. Because of this, it may be best to use them only as a specimen in their own selected spot.

WALNUT

"I went down into the garden of nuts to see the fruits of the valley, and to see whether the vine flourished, and the pomegranates budded." The Song of Solomon 6:11

Although this passage from The Song of Solomon only refers to "nuts" in the King James Version, you can trace further into other versions and find that Dr. Moffatt employs the phrase "walnut-bower." Most authorities now agree that this nut could indeed be the walnut, for several reasons.

The Persian or common walnut, *Juglans regia*, is generally believed to

Walnut trees most likely grew in Biblical times as they do in modern Israel today. They will provide you with their welcome shade and abundant crops of tasty nuts as part of your home landscape.

be a native of Persia. Although today it is found wild and cultivated from China to Europe and North America, the *regia* species has been mentioned in other writings that date back to Biblical days. Botanists believe that these walnuts were grown in Lebanon as well as Persia thousands of years ago and, indeed, in some parts of Gilead. Even today walnut trees that appear to be quite aged can be found, primarily in the higher elevations and near water sources. In Israel today wild walnut trees that were most likely started from some seeds of ancient walnut trees dropped many years ago are thriving.

Old writings from the Orient reveal that at about the time of Solomon, walnuts were grown in the far reaches of the world. Inasmuch as walnuts are easily transported, it is reasonable to expect that early travelers and merchants could have carried them from their native land to many other areas, including Palestine, as articles of trade.

The English walnut, *Juglans regia*, which is so widely grown in England, is actually the Persian walnut tree. This massive tree often reaches 100 feet high, spreading its branches wide from trunks that can be 6 feet thick on older trees. It prospers in the United States as well and is noted not only for its tasty nuts and ample shade but for its massive, dramatic appearance.

Walnut trees have been grown under cultivation for so many centuries in so many countries that it is difficult to trace their true origin. Although probably not native to the Holy Land, the scriptures do refer to these stately trees which most likely produced their nutritious rich, meaty nuts for the children of Israel. Now grown widely in Europe and Asia, walnuts probably arrived in England with the Romans. Plant breeders in England perfected improved strains, which accounts for the present name of "English" walnut. The cultivated trees yield more abundantly than the wildings and those that grew centuries ago even under cultivation.

You can grow walnuts on a tree in your yard as many others do. Only in California, however, has the warm climate lent itself to a substantial commercial production of these tasty nuts. For general home use, the English Carpathian strain of walnut has proved best. You should plant two trees to insure proper pollination. This Carpathian strain of *Juglans regia* grows 30 to 40 feet tall with a spreading head, graceful branches, and sweeping foliage.

Although the walnut trees of the Holy Land produced their crops in the typically hot climate, this new strain has been perfected to survive severe winter temperatures. They are hardy even in the extreme cold of northern states.

When you plant walnuts, prepare the soil as you would for other valuable trees, digging deeply and improving poor soil if necessary. To obtain large crops of nuts, water your trees regularly and deeply, since they need deep soil moisture. It is also important to keep the area around the trunk dry to avoid rot, which can form from excess water there. Walnut trees have a habit of feeding lower in the ground. A root-feeding device attached to your garden hose, such as a Ross root feeder, will let you combine fertilizer with water delivered 18 to 24 inches underground. During dry periods, you can also simply irrigate with water alone in a circle around the trees without nutrient pellets.

Walnuts, mentioned among the trees of the Bible, offer lovely shape and ample shade. Best of all, they provide their meaty nuts for picking in your own backyard.

WILLOW

"The shady trees cover him with their shadow; the willows of the brook compass him about." Job 40:22

"Therefore the abundance they have gotten, and that which they have laid up, shall they carry away to the brook of the willows." Isaiah 15:7

Willows and poplars are common trees of the Holy Land today as they were so many, many centuries ago. Although these trees are not especially valued for their wood, they have other virtues. Poplars grow rapidly and tall, providing dense shade from the ample branches so well covered with leaves. Willows also grow quickly from spindly saplings into reasonably tall trees. Neither could survive without ample water to supply their thirsty roots.

As you read the scriptures, compare if you wish the different translations—King James, Douay, Goodspeed, Moffatt, Jastrow—and you may be surprised to discover poplar and willow substituted from one version to the other. Since both these trees have similar growing habits, especially their need for enormous amounts of water to survive, they are found along stream banks and in wet or marshy areas. It seems only natural that confusion as to their identity arose in those days when the Bible was being written down and translated into different languages.

This constant need of water seems to have accurately identified the willow according to several passages. "And they shall spring up as among the grass, as willows by the water courses." Isaiah 44:4. In Ezekiel 17:5 you again find clear reference to water: "He took also of the seed of the land, and planted it in a fruitful field; he placed it by great waters, and set it as a willow tree." In Job 40:22, as you also read above, willows are associated with a brook.

Along brooks in the Holy Land, willow trees mark the course of water today as they did many centuries ago. You can grow graceful weeping willows if you have the moist ground they need to survive.

In the Douay and Jastrow versions of the Bible you will find willows in Psalms 137:2, as the King James Version does: "We hanged our harps upon the willows in the midst thereof."

In the debate between the willows and the poplars, some scholars add to the confusion by pointing to the oleander, *Nerium oleander.* It is more likely, despite the Goodspeed and Moffatt translations changing willows to poplars, that the willow was intended to be just that, a willow.

The willows *Salix acmophylla, Salix fragilis, Salix alba,* and *Salix safsaf* are believed to have been present in Palestine then as they are today. Willows are rather easily identified by their oblong, elliptical or lanceolate leaves, depending on the exact species. The bark of these trees, especially one-year-old new bark is usually light-colored, yellow, yellowish, or reddish-yellow. The trees have an airy appearance created by the slender, long leaves which give the trees, along with their tendency to have light, supple branches, a more graceful appearance than most other trees.

This slenderness of willow branches also leads many scholars to believe that willows were intended in the passages from Judges 16:7–9: "And Samson said unto her, If they bind me with seven green withs that were never dried, then shall I be weak, and be as another man. Then the lords of the Philistines brought up to her seven green withs which had not been dried, and she bound him with them. Now there were men lying in wait, abiding with her in the chamber. And she said unto him, The Philistines be upon thee, Samson. And he brake the withs, as a thread of tow is broken when it toucheth the fire. So his strength was not known."

It may sound odd to equate green withs with willows. However, traditionally, withs are slender, flexible branches, usually of willow wood, used in wicker weaving. With that meaning added from old texts, it is possible to agree that willows were the intended meaning in the passages from Judges.

Since willows are pleasant trees, and a wide choice of species native to America related to those from the Holy Land is readily available, they too deserve a place in your home landscape.

Where there are willows, there is water. In this respect, willows are much like date palms that mark the water of an oasis. As you travel across America, you can spot the waterways and even intermittent streams that dry up on the surface in the summer. Willows mark their path.

Willows have been clearly identified, of course, as one of the trees of the Bible. The context of the passages clearly seems to indicate that these fragile, often brittle trees were plentiful along brooks and swampy areas that did and do exist even in the arid areas of Israel. Although these trees are brittle, tending to lose limbs during wind and ice storms, their graceful flowing habit has won them many admirers. Weeping willows are conspicuous in the spring as their bark turns yellowish with the flow of new sap from the thawed ground.

You can grow willows anywhere you have reasonably fertile ground, but their first requirement is a moist environment. Willows require enormous amounts of water. Their roots roam in search of it. For this reason, never plant them near septic systems, leach fields, or sewer pipes. Like poplars, willow roots have a way of sneaking their way into these pipes, clogging them and causing costly problems. Another drawback is their tendency to drop leaves continually. More leaves form naturally, but cleaning up leaves in summer can be a chore.

Since this is established as one of the plants of the scriptures, you may wish to grow it. The graceful flowing branches that drape many feet from the limbs no doubt have a certain appeal.

The exact species which seem to be identified among botanists who have studied the scriptures include *Salix alba, fragilis,* and *acmophylla. Salix alba* has been improved by nurserymen and is now available in what appears as a golden form. It gets this name from the bright yellow color of the new twigs that form as it grows, especially during spring, when sap flows swiftly. The narrow, 4-inch-long leaves are bright green to yellow-green and are paler on their undersides. Mature willows may soar 60 to 70 feet high with a great ground spread. Around our family farm pond in New Jersey, two dozen ancient willows provided all the shade we needed to sneak away from haying chores to fish in the heat of summer.

Another recommended species is the Weeping willow, *S. babylonica.* This is a smaller version of the alba and its lower-growing size fits it more conveniently into restricted garden grounds. Graceful, especially drooping branches mark this tree well. You may find exact species of the Biblical plants from some nurseries, but most likely these relatives with the same growing habits will be more suitable for American gardens.

APPENDICES

A CHILD'S BIBLE GARDEN

Youngsters will appreciate God's world even more themselves when they can enjoy their own windowsill Bible garden. As they read the scriptures, learning about plants of the Bible, they can watch their own plants sprout, take root, grow tall, and burst into glorious blooms before their eyes.

It's easy to force bulbs of spring flowers into bloom in winter. Nothing says springtime better than these radiant bulb-flowers, with their individual scriptural significance. Your youngsters and their friends can watch colorful crocuses, sunshiny narcissi, fragrant hyacinths, and lovely tulips bloom inside your home in midwinter. If you teach Sunday School or help with classes at your church or synagogue, perhaps these ideas for blooming beauty from Biblical plants may be of extra value. As you read and teach the scriptures, your students can learn about the Holy Land and about the people and the lives they lived in Biblical times. Plants that come alive for youngsters can help make the scriptures come alive as well.

Indoor bulb gardening is both educational and good, old-fashioned fun for children. How often have we heard a young child ask, "How does a flower grow?" The extra beauty of forcing bulbs to bloom is that the process is virtually foolproof. The plants and their flowers are all preformed as growth buds inside the bulbs. With just a little water, warmth, and sun, the bulbs do all the work.

You can, of course, guide the youngsters as they plant the bulbs and add water. Older children can most likely take over and do the entire project by themselves with a few helpful hints from you. In other chapters of this book, you'll find the scriptural references about these plants. In this chapter, you'll find the simple how-to steps to blooming-bulb success indoors.

Crocuses, hyacinths, narcissi, and tulips can all be potted and forced to bloom indoors before their usual season. But not every variety of bulb can be grown successfully. Although the best varieties for forcing sometimes don't match the exact species from the scriptures, at least you can enjoy those that are similar in appearance and blooming habit. You usu-

ally will be able to find the desired, large-size bulbs that are best for forcing in local garden centers and florists. Often the stores will have complete
prepackaged kits. Reputable mail order firms such as Burpee, Parks, and
Wayside Gardens have excellent kits at reasonable prices, if you cannot
find what you need locally.

When you shop locally, choose bulbs that are plump, solid, and free
of deep scars. Bulb flowers grow in a delightful variety of shapes, sizes,
and colors, so plan to try several different plants. Youngsters can marvel
at the deep crimson of tulips with their classic cuplike blooms, blended
with the delicate pastel yellow of clusters of narcissus blooms and their
contrasting orange trumpets.

Timing is important to achieve the blooming period you desire. For
flowering in January, plant bulbs around October 1. For February flowering, plant in mid-October. For March and April blooms, pot bulbs in
mid-November. A brief list of the best varieties appears at the end of this
chapter. Those that most closely resemble the plants of the Bible are
starred.

Handle bulbs carefully at all times. Remember, they are living plants
and should not be dropped or subjected to extreme temperatures. If bulbs
are in bags when you buy them, open the bags to allow the bulbs to
breathe. If it is not convenient to pot bulbs immediately, keep them in a
cool room at a temperature of 55 to 60 degrees. At that temperature they
will remain dormant. Never let them freeze, of course.

You can use almost any kind of container for growing bulbs and forcing them to bloom indoors, but it must have a drainage hole at the bottom,
or some provision to let excess water drain away. The container should be
at least twice as high as the bulb to allow for adequate root development.
Garden centers usually have bulb pans or pots that are just the right size.
You can use more decorative containers if you wish. Ceramic bowls,
wooden baskets, clay or plastic pots all work well. If you use clay pots,
however, soak them overnight so they will not draw moisture from the
soil. Clay pots are porous and will breathe. That is an advantage, since
excess moisture can escape through the sides as well as out the bottom.

Flower bulbs prefer loose, crumbly soil. Try this simple test or let
the youngsters do it. Roll the soil into a tight ball. If it sticks together, it
is too heavy for your bulb plants. You can improve garden soil by adding
sand, peat moss, or vermiculite to loosen heavy soil. If possible, don't use
outdoor garden soil, which may have fungus and disease organisms in it.
The best mix of all is one part potting soil, one part peat moss, and one
part sand. For your project, a loose, open mixture is important. You won't
need any fertilizer, because all the food your plants require is already
stored in the bulbs.

Potting is the easiest part of all. Ask your youngster to place a small stone or pieces of broken clay pot over the drainage hole in the bottom of the empty pot. This prevents clogging and keeps the soil from flowing out. Then, fill the container with just enough of the soil mixture so that when the bulb is placed on top of the soil, its tip will just reach the rim of the container. Next, press the bulbs gently into the soil mix. Don't force them down, as this might damage the base of the bulbs.

You can place bulbs as closely as you like, so they almost touch one another if you prefer. For the fullest effect when they flower, use as many bulbs as will fit in the container neatly, with a half-inch space between them. Once you have your bulbs in place, fill the rest of the container with your planting mix. Be sure it settles between the bulbs. Pat the mix gently into place, but not too tightly. That will give bulbs snug comfort for a proper start. Then, cover the bulbs to within a half inch of the top of the container. The tips of your bulbs should just peek above the soil. Finally, water the container thoroughly.

The easiest way to water planted containers is to plunge them into a sink or large pan or pail of water about halfway up. Let the water soak up through the drainage hole until the top surface of your soil mix looks and feels moist. This method provides proper moisture throughout the container without disturbing the bulbs, which you could do by pouring water over them.

Be sure that you label each container with the variety name and color, and the date that they were planted. With only their tips peeking through, it is difficult to tell which bulbs are which.

All spring-flowering bulbs must have forced "winter cold" or dormancy in order to develop their roots. In the Holy Land, that dormancy is induced, of course, by the hot, dry season. In our climate, in the United States, bulbs go dormant in our winter. For forcing purposes, you must substitute this cold, dormant period.

Most bulbs require at least 13 weeks of cooling. Tulips and narcissi need 14 weeks, and some varieties may require up to 16 weeks of cold treatment. You can provide this necessary cold treatment by placing the containers indoors in a cold area, such as a northern corner of a basement or garage; or outdoors directly in the ground or in a convenient cold frame. You have a choice. Here's how to provide this important cooling period indoors or out.

Indoors, bulbs seem to root best in a cold, dark location under cover. An unheated garage, cold basement, or outdoor shed in which temperatures range between 40 and 45 degrees is desired. You should water your containers regularly when cooling them in such locations.

Outdoors, you should cool the bulbs in an outdoor pit or cold frame. This method, although a bit of work, is the best guarantee of successful flowering. Dig a bed that will hold all your containers. It should be about

an inch deeper than the tallest container. Set the pots close together and
water them well. Then, cover them with 6 inches of soil or sand. That's
all there is to that. The mound on the ground tells you where they are
when it is time to dig them up to begin forcing them inside.

The waiting period passes more quickly than you might expect. Providing you remembered to mark your calendars, you can begin reviving
your stored containers after the 14 to 16 weeks they need to set their roots.

Bringing the bulbs indoors to flower is, of course, the best part of the
project. The most enjoyable time is when they begin to burst into bloom.

Check your stored pots, those indoors as well as outdoors, between
the 12th and 13th week. Bulbs are ready to be moved into the house when
sprouts are well out of the bulb, about 2 or 3 inches high in most cases.
When the sprouts are at least 3 inches tall, it's time to move them into
their future home.

If you did your planting on October 1, your plants will be ready to
come indoors right after Christmas. That is the time when many blizzards
begin to blow. Your plants can help relieve that frigid feeling from your
outdoor view.

When you bring the plants in, place them in a cool room. The temperature should be 55 to 60 degrees at first. Let them stay in that room,
out of direct sunlight, for 2 or 3 weeks. You'll notice that they will be
growing faster during this period. When you see the buds begin to form,
you can bring them into their appointed blooming place. Continue to
water your containers whenever they feel dry. Within 3 to 4 weeks from
the time you brought them indoors, your crocuses, hyacinths, narcissi,
and tulips will be blazes of color and beauty. Not only are these plants a
thrill to see, but you and your youngsters will have learned more of God's
wonders with this project.

Then, while winter winds blow and snow drifts outside your home,
your whole family, friends and neighbors and their children too, can marvel at these plants of the Bible on your windowsill.

Try these bulbs which are recommended for forcing. The color of
each bulb's blossoms is listed on the right.

TULIPS

Christmas Marvel	Cherry-pink
Thule	Red, edged yellow
Paul Richter	Red
Trance	Deep geranium red, base yellow
Kees Nelis	Blood red, edged orange-yellow
Prominence	Dark red
Danton	Deep carmine
Golden Eddy	Red, edged yellow

NARCISSI

·Dutch Master	Yellow
Flower Record	White, yellow cup, edged orange
Fortune	Yellow, soft orange cup
Ice Follies	White two-tone

MINOR BULBS

Crocus	Yellow, white, blue, violet, striped
Crocus species	Yellow, white, blue, violet, striped
Iris reticulata	Violet

HYACINTHS

Ostara	Pansy violet with light purple blue
Amsterdam	Phlox purple with darker midrib
Blue Jacket	Dark blue, purple striped

This is a selective list. Other types of bulbs can be forced. So, if you have a particular favorite that is not on this list, don't be afraid to experiment. It may very well prove excellent for forcing.

Then, when the icy winds are howling, you can curl up by the fireside and enjoy your own private spring garden, indoors, reminiscent of these plants of the Bible.

SEED, ROOT, BULB, P·LANT SOURCES

As you dig more deeply into the good earth and begin cultivating greater interest in plants of the Bible, you may find some plants available locally. The more common trees, flowers, and vegetables and their direct descendants may be available from garden centers and florists.

During research for this book, we discovered many more seeds and plants in mail order catalogs. Not all of the firms listed here have all the plants covered in this book. However, you can find many of them, from fruit trees in the Starks catalog to cyclamens and narcissi in the Parks and Burpee catalogs. Most catalogs are free. You can obtain them by writing to these firms:

W. Atlee Burpee Company 300 Park Avenue Warminster, PA 18974	Flowers, vegetable, herb seeds, trees, plants, bulbs
Wayside Gardens Hodges, SC 29653	Trees, shrubs, plants
Park Seed Company Greenwood, SC 29646	Flowers, vegetable, herb seeds, trees, plants, bulbs
Stark Bros. Nurseries Louisiana, MO 63353	Fruit trees, grapevines
Bountiful Ridge Nurseries Princess Anne, MD 21853	Fruit trees, bushes, grapevines
Burgess Seed and Plant Co. Galesburg, MI 49053	Seeds, fruit plants, bulbs
Farmer Seed and Nursery Co. Faribault, MN 55021	Seeds, fruit plants, bulbs
Henry Field Seed & Nursery Co. Shenandoah, IA 51601	Seeds, fruit plants, bulbs
Gurney Seed & Nursery Co. Yankton, SD 57078	Seeds, fruit plants, bulbs
Inter-State Nurseries, Inc. Hamburg, IA 51640	Fruit trees, grapevines, trees
R. H. Shumway Rockford, IL 61101	Seeds, plants, bulbs
Armstrong Nurseries P.O. Box 473 Ontario, CA 91761	Trees
Stribling's Nurseries P.O. Box 793 Merced, CA 95340	Trees

HORTICULTURAL LIBRARIES

Many libraries have good references about plants. However, if you wish to pursue more extensive studies of plants of the Bible in depth, your best bet for detailed information and a wide range of source books is a specialized horticultural library. There are many around the country, at colleges and arboretums. Here is a list of the major horticultural libraries for your convenience.

California

California Academy of Sciences Library
Golden Gate Park
San Francisco, CA 94118

California State Polytechnic College Library
3801 West Temple Avenue
Pomona, CA 91768

Forest History Society
Box 1581
Santa Cruz, CA 95061

Los Angeles State and County Arboretum Library
301 North Baldwin Avenue
Arcadia, CA 91006

Rancho Santa Ana Botanic Garden Library
1500 North College Avenue
Claremont, CA 91711

Colorado

Denver Botanic Gardens
Helen K. Fowler Library
909 York Street
Denver, CO 80209

District of Columbia

Dumbarton Oaks Garden Library
1703 32nd Street N.W.
Washington, DC 20007

U. S. National Arboretum Library
U. S. National Arboretum
Washington, DC 20002

Florida

Fairchild Tropical Garden
Montgomery Library
10901 Old Cutler Road
Miami, FL 33156

Hume Library
University of Florida
Gainesville, FL 32611

Georgia

Callaway Gardens
Pine Mountain, GA 31822

Illinois

Chicago Horticultural Society Library
116 S. Michigan Avenue
Chicago, IL 60603

Lake Forest Library
360 Deerpath Avenue
Lake Forest, IL 60045

Morton Arboretum
Sterling Morton Library
Lisle, IL 60532

Indiana

Purdue University
Forestry-Horticulture Library
Lafayette, IN 47907

Maryland

National Agricultural Library
U. S. Department of Agriculture
Intersection I-495 and U. S. 1
Beltsville, MD 20705

Massachusetts

Arnold Arboretum
The Arborway
Jamaica Plain, MA 02130

Massachusetts Horticultural Society
300 Massachusetts Avenue
Boston, MA 02115

Oakes Ames Orchid Library
22 Divinity Avenue, Room 109
Cambridge, MA 02138

Old Sturbridge Village Library
Sturbridge, MA 01566

University of Massachusetts
Morrill Library
Amherst, MA 01002

Wellesley College Library
Sage Hall, Wellesley College
Wellesley, MA 02181

Worcester County Horticultural Society
30 Elm Street
Worcester, MA 01609

Michigan

Michigan Horticultural Society
The White House, Belle Isle
Detroit, MI 48207

Minnesota

University of Minnesota
St. Paul Campus Library
St. Paul, MN 55101

Mississippi

Mississippi Agricultural Experiment Station
Stoneville, MS 38776

Missouri

Missouri Botanical Garden
2315 Tower Grove Avenue
St. Louis, MO 63110

National Council of State Garden Clubs, Inc.
4401 Magnolia Avenue
St. Louis, MO 63110

New Hampshire

University of New Hampshire Biological Sciences Library
Kendall Hall
Durham, NH 03824

New Jersey

Rutgers University
College of Agriculture and Environmental Science
New Brunswick, NJ 08903

New York

Cornell University
Albert R. Mann Library
Ithaca, NY 14850

Garden Center of Rochester
5 Castle Park
Rochester, NY 14620

Highland Park Herbarium Library
Monroe County Parks
375 Westfall Road
Rochester, NY 14620

Horticultural Society of New York, Inc.
128 West 58 Street
New York, NY 10019

New York Botanical Garden Library
New York Botanical Garden
Bronx, NY 10458

New York State Agricultural Experiment Station Library
Geneva, NY 14456

State University of New York
Walter C. Hinkle Memorial Library
Alfred, NY 14802

North Carolina

North Carolina State University
D. H. Hill Library
Raleigh, NC 27607

University of North Carolina
Botany Library
301 Coker Hall
Chapel Hill, NC 27514

Ohio

Garden Center of Greater Cleveland
Eleanor Squire Library
11030 East Boulevard
Cleveland, OH 44106

Holden Arboretum Library
9500 Sperry Road
Kirtland P.O.
Mentor, OH 44060

Kingwood Center Library
Box 1186
Mansfield, OH 44901

Ohio Agricultural Research and Development
Center Library
Wooster, OH 44691

Youngstown Garden Center
123 McKinley Avenue
Youngstown, OH 44509

Oregon

Oregon State University Library
Oregon State University
Corvallis, OR 97330

Pennsylvania

Hunt Botanical Library
Carnegie-Mellon University
Pittsburgh, PA 15213

Longwood Gardens Library
Longwood Gardens
Kennett Square, PA 19348

Morris Arboretum
University of Pennsylvania
9414 Meadowbrook Lane
Philadelphia, PA 19118

Pennsylvania Horticultural Society
325 Walnut Street
Philadelphia, PA 19106

Pennsylvania State University
Agricultural Sciences Library
University Park, PA 16802

Temple University
Ambler Campus Library
Meetinghouse Road
Ambler, PA 19002

South Dakota

South Dakota State University
Lincoln Memorial Library
Brookings, SD 57007

Texas

Texas A&M University
Box 236
Weslaco, TX 78596

Vermont

University of Vermont
Guy W. Bailey Library
Burlington, VT 05401

Washington

University of Washington Arboretum
University of Washington
Seattle, WA 98105

West Virginia

Wheeling Garden Center Library
Oglebay Park
Wheeling, WV 26003

Wyoming

University of Wyoming
Coe Library
Laramie, WY 82070

Canada

Civic Garden Centre Library
777 Lawrence Avenue E.
Don Mills, Ontario, Canada

Royal Botanical Gardens
Box 399, Station A
Hamilton, Ontario, Canada

BIBLIOGRAPHY

Through the centuries Biblical scholars, botanists, students of natural history and horticulture have explored the scriptures. There are hundreds of books and articles about plants of the Bible that you may consult as you pursue your own research into this fascinating field.

The Bible itself, in its various editions and translations, remains our basic reference, of course. However, among the many books, these may prove most helpful in furthering your knowledge.

Alon, A. *The Natural History of the Land of the Bible*, 1969, 276 pages

Balfour, J. H., *The Plants of the Bible; Trees and Shrubs*, London, 1857

Balfour, J. H., enlarged edition of 250 pages, 1885

Cotes, R. A., *Bible Flowers*, 1904, 288 pages

Crowfoot, G. M., and L. Baldensperger, *From Cedar to Hyssop: A Study in the Folklore of Plants in Palestine*, 1904, 204 pages

Harris, T. M., *The Natural History of the Bible*, 1824, 462 pages

Hastings, G. T., *Plants of the Bible, a Review in Torreya*, 1942

Moldenke, H. N., *Plants of the Bible*, 1940, 135 pages

Moldenke, H. N. and A. L. E., *Plants of the Bible*, 1952, 328 pages

Temple, A. A., *Flowers and Trees of Palestine*, 1907, 184 pages

Untermeyer, L., *Plants of the Bible*, 1970, 26 pages

Among the many magazines and periodicals issued by religious orders and organizations, as well as for the lay person, and among mass-circulation publications, you can find hundreds of different articles about Biblical plants. Many are excellent reading but often contain much of the same conjectures that have puzzled botanists for years.

Nevertheless, if you wish to pursue your own investigations in this growing field of plants of the Bible, your local and regional libraries can provide guidance to these many periodicals.

Most likely, the best two references are Cruden's Concordance, which guides you to the precise scriptural passages, and Moldenke's *Plants of the Bible*, with its detailed botanical information.

Since local libraries may not have or be able to easily obtain the reference books you wish to read, you may find it helpful to visit one of the many Horticultural Libraries in the United States. These libraries specialize in books and other writings concerned with the field of horticulture. A list of these key libraries is included here. Theological seminaries and other religious institutions, from colleges to divinity schools, are also potential reference sources, but their number precludes listing in this book.

ABOUT THE AUTHOR

ALLAN A. SWENSON, author of more than a half dozen gardening books, is Director of the Book Division, Guy Gannett Publishing Company, in Portland, Maine. His articles as garden columnist for the Newspaper Enterprise Association since its inception 20 years ago are featured in some 300 newspapers nationwide each year. In 1966, Mr. Swenson launched a Gardeners Notebook radio show which was featured on both the Mutual Radio System and selected independent stations.

Mr. Swenson graduated Phi Beta Kappa from Rutgers University, where he majored in journalism. He had produced numerous syndicated radio programs and TV specials and is presently continuing these activities from his studios in Maine, where he, his wife Sheila, and their four children reside.

INDEX

Absinthe, 111

Acacia tree *(Acacia)*, 6, 154-58; *A. baileyana* (Cootamundra Wattle), 158; *A. decurrens* (Green Wattle), 158; *A. pendula* (Weeping Myall), 158; *A. pruinosa*, 158; *A. seyal*, 157; and ark of the tabernacle, 154-56; *A. Tortilis*, 157; Bible references and identification, 154-57; and charcoal, 157; and drought resistance, 156; flowers, 158; growing, 157-58; gum arabic from, 157; as shittah tree and shittim wood, 154-56, 157; species, 156, 158

Adam and Eve, xvi, 120. *See also* Garden of Eden

Almond tree *(Prunus amygdalus)*, 6, 36, 154, 158-61; and Aaron's rod, 159, 161; Bible references, 158-59, 160; bitter almonds from, 160; blossoms, 160; branches, 159; fruit, 160; growing, 160; leaves, 160; Nonpareil, 160; nuts (almonds) from, 158-61; planting and growing, 160-61; soil for, 160; sweet almonds from, 160; Texas, 160; wood, 159

Aloe, 6, 99-100; *Aloe vera*, 100; *Aloe vera chinensis*, 100; American *(Agave americana)*, 99; Bible references and identification, 36, 96, 99; century, 99; flowers, 100; growing, 99-100; medicinal uses, 100; Spiny *(A. africana)*, 100; Tree *(A. arborescens)*, 100; true *(A. succotrina)*, 99-100

Alpine violet, 51. *See also* Cyclamen

Amos, 163, 172

Anemone *(Anemone coronaria)*, 44-48; Bible references and identification, xx, 12, 13, 34, 44-47; container gardening, 26, 47; flowers, 47; as lily of the field, xx, 44-47; planting and growing, 47-48; as poppy anemone, 47; soil for, 47; as windflower, 47

Anise, *(Pimpinella anisum)*, 101, 102

Apples (apple tree), 6, 27, 34, 116, 117-23; "apples of gold," 118; and apricots, 117-23; Bible references and identification, 27, 117-20; climate and, 118; common *(Malus pumila)*, 117-18; Golden Delicious, 122; home growing, 26, 119-23; Lodi, 123; Northern Spy, 123; Ozark Gold, 122; pest control, 123; planting and growing, 119-23; Red Boquet Delicious, 123; Starkspur McIntosh, 122; Starkspur Red Rome Beauty, 122; and "tree of life," 117; Tropical Beauty, 122; varieties, 122

Apricot *(Prunus armeniaca)*, 6, 27, 113, 117-23; apples and, 117-23; Bible references and identification, 117-20; Chinese Golden, 119; climate for, 119; Earli-Orange, 119; flowers, 118-19, 120; as "golden apples," 118; home growing, 17, 26, 119-23; Hungarian Rose, 119; Moorpark, 119; pest control, 123; planting and growing, 119-23; soil for, 119-20; Stark Giant Tilton, 119; varieties, 119; Wilson Delicious, 119

Arabian Desert, 11

Arbatus, 13

Arboretums, libraries of (listed), 197-203

Armstrong Nurseries, address of, 196

Army worms, 116. *See also* Caterpillars

Ashes, wood, as a source of potash, 19